Testing for Small-Delay Defects in Nanoscale CMOS Integrated Circuits

Devices, Circuits, and Systems

Series Editor
Krzysztof Iniewski
CMOS Emerging Technologies Research Inc.,
Vancouver, British Columbia, Canada

FORTHCOMING TITLES:

Labs-on-Chip: Physics, Design and Technology
Eugenio Iannone

Laser-Based Optical Detection of Explosives
Paul M. Pellegrino, Ellen L. Holthoff, and Mikella E. Farrell

Metallic Spintronic Devices
Xiaobin Wang

Microfluidics and Nanotechnology: Biosensing to the Single Molecule Limit
Eric Lagally and Krzysztof Iniewski

**MIMO Power Line Communications: Narrow and Broadband Standards,
EMC, and Advanced Processing**
Lars Torsten Berger, Andreas Schwager, Pascal Pagani, and Daniel Schneider

Mobile Point-of-Care Monitors and Diagnostic Device Design
Walter Karlen and Krzysztof Iniewski

Nanoelectronics: Devices, Circuits, and Systems
Nikos Konofaos

Nanomaterials: A Guide to Fabrication and Applications
Gordon Harling and Krzysztof Iniewski

Nanopatterning and Nanoscale Devices for Biological Applications
Krzysztof Iniewski and Seila Selimovic

Optical Fiber Sensors and Applications
Ginu Rajan and Krzysztof Iniewski

Power Management Integrated Circuits and Technologies
Mona M. Hella and Krzysztof Iniewski

Radio Frequency Integrated Circuit Design
Sebastian Magierowski

Semiconductor Device Technology: Silicon and Materials
Tomasz Brozek and Krzysztof Iniewski

Smart Grids: Clouds, Communications, Open Source, and Automation
David Bakken and Krzysztof Iniewski

Solar Cells: Materials, Devices, and Systems
Qiquan Qiao and Krzysztof Iniewski

Soft Errors: From Particles to Circuits
Jean-Luc Autran and Daniela Munteanu

Technologies for Smart Sensors and Sensor Fusion
Kevin Yallup and Krzysztof Iniewski

VLSI: Circuits for Emerging Applications
Tomasz Wojcicki and Krzysztof Iniewski

Wireless Transceiver Circuits: System Perspectives and Design Aspects
Woogeun Rhee and Krzysztof Iniewski

Testing for Small-Delay Defects in Nanoscale CMOS Integrated Circuits

Edited by
Sandeep K. Goel
Krishnendu Chakrabarty

CRC Press
Taylor & Francis Group
Boca Raton London New York

CRC Press is an imprint of the
Taylor & Francis Group, an **informa** business

CRC Press
Taylor & Francis Group
6000 Broken Sound Parkway NW, Suite 300
Boca Raton, FL 33487-2742

First issued in paperback 2017

© 2014 by Taylor & Francis Group, LLC
CRC Press is an imprint of Taylor & Francis Group, an Informa business

No claim to original U.S. Government works

Version Date: 20130801

ISBN 13: 978-1-138-07577-1 (pbk)
ISBN 13: 978-1-4398-2941-7 (hbk)

Library of Congress Cataloging-in-Publication Data

Testing for small-delay defects in nanoscale CMOS integrated circuits / editors, Sandeep K. Goel, Krishnendu Chakrabarty.
    pages cm. -- (Devices, circuits, and systems)
    Includes bibliographical references and index.
    ISBN 978-1-4398-2941-7 (hardcover : alk. paper)
    1. Metal oxide semiconductors, Complementary--Testing. I. Goel, Sandeep K., editor of compilation. II. Chakrabarty, Krishnendu, editor of compilation.

TK7871.99.M44T43 2014
621.39'732--dc23   2013028538

Visit the Taylor & Francis Web site at
http://www.taylorandfrancis.com

and the CRC Press Web site at
http://www.crcpress.com

Contents

Section IV SDD Metrics

Preface

Recent years have seen tremendous advances in design methods and process technologies in the semiconductor industry. These advances have resulted in a relentless increase in the complexity of integrated circuits (ICs). Due to high complexity and nanometer feature sizes, today's ICs are not only susceptible to manufacturing defects such as resistive opens and shorts, but also affected by process variations, power-supply noise, cross talk, and rule violations related to design for manufacturability (DfM), such as butted contacts and insufficient via enclosures. These defects and abnormalities introduce small delays in the circuit, collectively referred to as small-delay defects (SDDs). Concern about SDDs in the semiconductor industry is highlighted by the fact that they cause immediate failures if they occur on critical paths in the circuit, and even if they occur on noncritical paths, they pose a quality risk and potential reliability hazard in the field. Due to these concerns, testing for SDDs is important in the latest technologies to achieve high product quality, measured in terms of low defective-parts-per-million (DPPM) rates.

Commonly used fault models such as single stuck-at and transition delay faults provide adequate coverage of most manufacturing defects, but they are not sufficient to ensure low DPPM in the presence of SDDs for high-performance chips. While the stuck-at fault model is likely to miss delay defects, the transition fault model is capable of screening gross delay defects. However, transition fault testing is not effective for SDDs because test generation using this fault model does not take into account important timing information needed for SDD screening; hence, it is more likely to generate patterns that detect faults only on short paths. The detection of SDDs requires fault activation and propagation through the longest paths in the circuit.

The path delay fault model can be used to target SDDs, but it requires enumeration of all sensitizable paths in the design. Since the number of paths increases exponentially with design size, the use of the path delay fault model for testing of SDDs is impractical for large ICs. Non-timing-aware transition fault patterns can also be used to detect SDDs if they are applied at a clock speed higher than the system clock frequency. Although these "faster-than-at-speed" patterns have been effective in detecting SDDs, chances of hazards and higher power consumption during tests due to higher clock speed can cause unexplainable failures and unnecessary yield loss (overkill) due to the scrapping of good parts.

It is, therefore, not surprising that SDD testing has been a subject of vigorous research by leading groups in universities and industry. A large number of breakthrough papers have been published recently in leading journals and conference proceedings of the test technology field. Timing-aware automatic test pattern generation (ATPG) tools from commercial electronic design automation (EDA) companies take delay information from the standard delay format (SDF) files into account for each transition fault and then attempt to excite and observe fault effect along the longest path. Such an approach detects a wider range of delay defects while the circuit is operated at the rated clock speed. However, when these ATPG tools are run in timing-aware mode, the resulting test pattern count and tool run time increase significantly; hence, they may not be practical. Therefore, efficient methods that reduce the pattern count while achieving similar or higher delay test coverage in less run time are sought in the semiconductor industry.

This edited book provides a comprehensive analysis of SDD testing challenges and practical solutions. The book provides in-depth analysis of the effectiveness and drawbacks of all state-of-the-art methods. It serves as a single source of information on this topic and highlights academic research as well as practical industry solutions and current practice. It includes ten chapters that have been divided into four parts.

Chapter 1, coauthored by veteran university researcher S. M. Reddy and industry pioneer Peter Maxwell, provides an overview of semiconductor industry test challenges and the need for SDD testing. Basic concepts and introductory material are included. This chapter leads to Section I of the book, on timing-aware ATPG. This part includes two chapters; the first is contributed by Duncan (Hank) Walker, and the second is from a leading team from Mentor Graphics Corporation. Readers will find a description of algorithmic solutions that have been incorporated in commercial tools from Mentor Graphics.

Section II of the book is on at-speed testing. It includes two chapters contributed by Mohammad Tehranipoor and his students/collaborators. Section III of the book describes SDD testing based on "alternative methods" that explore new metrics, top-off ATPG, and circuit topology-based solutions. This part includes three chapters from two leading companies (AMD and LSI Corporation) and Duke University. Section IV of the book, "SDD Metrics," presents various methods that have been advocated to evaluate the SDD coverage in a quantitative way. The chapter highlights the advantages and disadvantages of a diverse set of metrics and identifies scope for improvement.

We hope that this book is stimulating and informative. We have included contributions from three viewpoints: university researchers, EDA tool developers, and chip designers and tool users. This book is the first of its kind to address all aspects of SDD testing from such a diverse perspective. We expect the book to serve as a one-stop reference for current industrial practices, research challenges in the domain of SDD testing, and some of the most compelling ideas and solutions. We thank the authors for their contributions and the staff at CRC Press for their encouragement and support. Finally, we acknowledge the financial support received in the form of research grants and contracts from the National Science Foundation and the Semiconductor Research Corporation.

Sandeep K. Goel
TSMC
San Jose, CA

Krishnendu Chakrabarty
Duke University
Durham, NC
June 2012

About the Editors

Sandeep Kumar Goel is a senior manager (DFT/3D-Test) with Taiwan Semiconductor Manufacturing Co. (TSMC), San Jose, CA. He received his PhD degree from the University of Twente, The Netherlands. Prior to TSMC, he was in various research and management positions with LSI Corporation CA, Magma Design Automation, CA, and Philips Research, The Netherlands. He has co-authored two books, three book chapters, and published over 80 papers in journals and conference/workshop proceedings. He has delivered several invited talks and has been panelist at several conferences. He holds fifteen US and five European patents and has more than thirty other patents pending. His current research interests include all topics in the domain of testing, diagnosis, and failure analysis of 2D/3D chips. Dr. Goel was a recipient of the Most Significant Paper Award at the IEEE International Test Conference in 2010. He serves on various conference committees including DATE, ETS, ITC, DATA, and 3DTest. He was the general chair of 3D Workshop at DATE 2012. He is a senior member of the IEEE.

Krishnendu Chakrabarty received the BTech degree from the Indian Institute of Technology, Kharagpur, in 1990, and MSE and PhD degrees from the University of Michigan, Ann Arbor, in 1992 and 1995, respectively. He is now professor of electrical and computer engineering at Duke University. He is also a chair professor in software theory at Tsinghua University, Beijing, China; a visiting chair professor of computer science and information engineering at National Cheng Kung University in Taiwan; and a guest professor at the University of Bremen in Germany. Professor Chakrabarty is a recipient of the National Science Foundation Early Faculty (CAREER) award; the Office of Naval Research Young Investigator award; the Humboldt Research Fellowship from the Alexander von Humboldt Foundation, Germany; and several best paper awards at IEEE conferences.

Professor Chakrabarty's current research projects include testing and design for testability of integrated circuits; digital microfluidics, biochips, and cyberphysical systems; and optimization of digital print and enterprise systems. In the recent past, he has also led projects on wireless sensor networks, embedded systems, and real-time operating systems. He has authored twelve books on these topics, published more than 420 papers in journals and refereed conference proceedings; and given more than 185 invited, keynote, and plenary talks. He has also presented twenty-five tutorials at major international conferences. Professor Chakrabarty is a Fellow of IEEE, a Golden Core Member of the IEEE Computer Society, and a Distinguished Engineer of ACM. He was a 2009 Invitational Fellow of the Japan Society for the Promotion of Science (JSPS). He is a recipient of the 2008 Duke University Graduate School Dean's Award for excellence in mentoring and the 2010 Capers and Marion McDonald Award for Excellence in Mentoring and Advising, Pratt School of Engineering, Duke University. He served as a distinguished visitor of the IEEE Computer Society during 2005–2007 and as a distinguished lecturer of the IEEE Circuits and Systems Society during 2006–2007. Currently, he serves as an ACM distinguished speaker; he served as a distinguished visitor of the IEEE Computer Society for 2010–2012 and a distinguished lecturer of the IEEE Circuits and Systems Society (2012–2013).

Professor Chakrabarty is the editor in chief of *IEEE Design and Test of Computers* and *ACM Journal on Emerging Technologies in Computing Systems*. He is also an associate editor of *IEEE Transactions on Computer-Aided Design of Integrated Circuits and Systems, IEEE Transactions on Circuits and Systems II,* and *IEEE Transactions on Biomedical Circuits and Systems*. He serves as an editor of the *Journal of Electronic Testing: Theory and Applications* (JETTA). In the recent past, he has served as associate editor of *IEEE Transactions on VLSI Systems* and *IEEE Transactions on Computer-Aided Design of Integrated Circuits and Systems,* and *IEEE Transactions on Circuits and Systems I.*

Contributors

Nisar Ahmed
Freescale Semiconductor
Austin, Texas, USA

Narendra Devta-Prasanna
LSI Corporation
San Francisco, California, USA

Mark Kassab
Mentor Graphics
Portland, Oregon, USA

Xijiang Lin
Mentor Graphics
Portland, Oregon, USA

Peter Maxwell
Aptina Imaging
San Jose, California, USA

Benoit Nadeau-Dostie
Mentor Graphics
Ottawa, Canada

Ke Peng
Freescale Semiconductor
Austin, Texas, USA

Sudhakar Reddy
College of Engineering
University of Iowa
Iowa City, Iowa, USA

Mohammad Tehranipoor
Department of Electrical and
    Computer Engineering
University of Connecticut
Storrs, Connecticut, USA

Hank Walker
Department of Computer Science
    and Engineering
Texas A&M University
College Station, Texas, USA

Mahmut Yilmaz
NVIDIA
San Francisco, California, USA

1

Fundamentals of Small-Delay Defect Testing

Sudhakar M. Reddy and Peter Maxwell

CONTENTS

1.1 Introduction

Test of very large scale integration (VLSI) devices is essential to ensure the quality of shipped products. With decreasing feature sizes in VLSI devices, the accuracy of device and interconnect models is diminishing. In addition, process variations are introducing greater variations in devices across wafers and dies. Systematic defects and design-specific defects are becoming more common. For this reason, tests and diagnosis are expected to play an increasing role in identifying yield- and reliability-limiting effects necessary for yield learning. In this chapter, fault models used to generate tests for delay defects are discussed.

1.2 Trends and Challenges in Semiconductor Manufacturing

1.2.1 Process Complexity

Modern semiconductor fabrication processes are extremely complex. In the early days, a complementary metal oxide semiconductor (CMOS) 5-μm process typically involved nine mask levels and around 100 steps [Smith 1997]. As the number of metal layers increased and geometries decreased, so did process complexity. A typical 0.25-μm process had over 20 layers, which has increased for present-day mainstream processes into the 30s [ITRS 2007], with over 1,000 process steps. For some specialized applications, such as optical sensor chips, a significant number of additional mask layers and steps are required.

The number of design rules has also increased dramatically. Much of the explosion in number is due to printing challenges of feature sizes, which are significantly smaller than the 193-nm wavelength light source used for patterning. Due to distortions in the projection system with such fine features, the shapes on a mask are not what are printed on silicon, and unless corrections are made, design rule violations could occur with minimum width and spacing. So-called reticule enhancement techniques (RETs) are needed to address the problem, involving predistorting mask patterns using optical proximity correction (OPC) algorithms so the final printed pattern reproduces the intended one accurately. This first became necessary at the 180-nm node, and as nodes progress to smaller and smaller geometries, increasingly sophisticated methods have to be utilized [Braun 2005; Raina 2006]. As printed line widths become smaller, design rules become increasingly restrictive in what are allowable layout patterns, and an increasing number of them are required.

Figure 1.1 shows the growth in design rules [Hess 2010]; it illustrates the complexity of modern processes. The implications are twofold. First, every

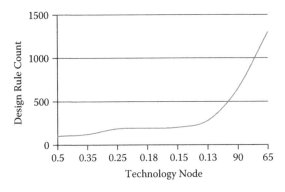

FIGURE 1.1
Number of design rules with technology node.

additional mask or step represents the possibility of some form of contamination being introduced that will affect circuit operation. Even if the process steps are for some specialized circuitry, such as a sensor array, the additional steps open the possibility of defects affecting all the circuitry, not just that for which the steps are required.

The second implication is that, with such a large number of design rules, there is little tolerance of small disturbances. Defects can occur that are dependent on such aspects as pattern proximity, pattern density, and less-than-perfect optical proximity correction algorithms. The more complex the rules are that relate to such factors, the more likely small perturbations are to cause a problem.

1.2.2 Process Variability

Decreasing geometries have resulted in more than additional complexity. Device parameters are determined by the physical way in which the devices are formed during fabrication. Their behavior is governed by a number of physical attributes, including size, shape, orientation, and dopant levels. The ability to control variations is being increasingly challenged. A representative 65-nm transistor has a gate oxide thickness of 1.3 nm. Given that a molecule of SiO_2 is about 0.3 nm the oxide thickness consists of only four or five molecules. A difference of one molecule represents a 20% change. This will translate to a change in threshold voltage V_t, subthreshold leakage current I_{subth}, and active current I_{DS}. Similarly, the average number of dopant atoms in the channel of a 32-nm device is less than 100 [Kuhn 2008], and random fluctuations in this small number can produce large variations in threshold voltage.

Major sources of process variability are as follows:

- Lithography, which affects channel dimensions and interconnect (affects I_{on}, I_{off}, V_t)
- Gate oxide thickness variability (affects I_{on}, I_{off}, V_t through C_{ox})
- Doping concentrations and profiles (affects V_t, channel length, body effect)

Such effects translate into significant intradie parameter variations. For example, one study showed that in a 65-nm process, channel doping variation gave rise to up to 30% change in threshold voltage. In turn, such a spread in variation can result in a 5% to 25% spread in the clock-to-output delay of a flip-flop depending on data arrival time [Mahmoodi 2005]. Additional complications arise because variations are layout dependent, with different orientations and shape of a device having significantly different parameters [Segura 2002].

Although the causes of variability are understood, predicting the results on silicon is a serious challenge. Designing for the worst-case conditions is unrealistic as it would lead to reduced performance and excessive area and power consumption. To address these concerns, there has been much research in statistical-based design, both at the gate level [Srivastava 2005] and at a high level [Xie 2009]. There are also a number of design practices and layout restrictions that can mitigate the effects of variability [Kuhn 2008], but even with the most sophisticated approaches, the models employed are incapable of exactly representing silicon performance. In some cases, postsilicon tuning is used to ensure correctly operating designs [Meijer 2012; Kulkarni 2006], but there is always the possibility of variation being outside either allowable tuning or model predictions.

Since variation has a continuous distribution, there will inevitably be situations when parameter changes produce only a small difference from allowable extremes. Such differences could be in any of a number of specifications, but we are particularly interested in circuit delay. The variations discussed can produce small additional delays, causing malfunction, which must be detected for high-quality manufacturing.

Although timing tools are supposed to predict path timing and allow robust designs, the reality is they are unable to handle all the possible process variations coupled with variations in circuit activity, resulting in the inability to always identify failing paths in silicon. Some authors [Killpack 2008] showed effects such as multiple input switching, noise cross-coupling, and localized voltage droop can cause unexpected paths to fail. The severity of the effect can clearly vary, but it is evident that small additional delays are likely to be produced.

1.2.3 Random versus Systematic Defects

As discussed, modern CMOS processes use geometries that are significantly smaller than the resolution of the lithographic equipment used to print patterns on silicon. The subwavelength lithography shifted the balance of causes of defective circuits.

There are two main classes of defects that cause a circuit to fail. One is due to visible physical defects, which deform structures on silicon and produce such irregularities as open interconnect, missing vias, transistor shorts, and so on. These sorts of defects are random in nature and were the prevalent type in early CMOS. The term *random defects* is often used for this type of defect but is somewhat of a misnomer because random effects can also influence the other main type of defect, which is parametric [Segura 2002]. Parametric failures result from electrical variations in the behavior of transistors, interconnect, and vias. Unlike physical defects, the variations are not visible, and many fall into the class of systematic defects. Examples are stress in vias, lithography focus or misalignment, copper dishing, and metal width/thickness variations.

Although many parametric defects are systematic, a significant number are, in fact, random. One example has already been given, namely, random dopant fluctuation. Others are line-edge and line-width roughness, gate dielectric, oxide thickness, and fixed charge. Ultimately, all these are due to physical causes, but the effects on the circuit are parametric variation. These parametric variations in turn can give rise to circuit fails. Table 1.1 [Segura 2002] shows a number of varied effects from different physical mechanisms. Notable is that the vast majority have an impact on delay. Although some of these will not result in circuit failure, many will, meaning a solid delay test strategy is imperative.

1.2.4 Implications of Power and Timing Optimization

A logic network typically has a large number of paths through it, and the delays along the paths form some distribution. It has long been known that timing optimization can be achieved by tightening up this distribution [Williams 1991]. Employing smaller drivers on short paths reduces both power and area requirements, both of which are critical constraints in modern designs. Nothing is to be gained by having unnecessarily large drivers that produce short paths. Lengthening what would otherwise be short paths changes the delay distribution as shown in Figure 1.2. Figure 1.2a shows a non-optimized distribution, which is transformed into one shown in Figure 1.2b by the optimization process. If this process could be done perfectly, all paths would have the same length and therefore the same slack with respect to the system clock. In such an ideal situation, a relatively simple test strategy could be employed since the detection of delay defects becomes independent of the path along which the testing is done [Park 1991].

In reality, perfect matching of delays is not feasible; in addition, process variation causes variation in delay, which will widen the distribution. Further, this variation will not be the same for the same path on other dies, which means that different dies can have different critical paths. If a path that would not normally be considered critical has increased delay to the point at which the timing slack is extremely small, it is clear that a small-delay defect could cause that path to fail on a particular die. The same defect on the same path in another die may give no problem, which makes it extremely difficult for design tools to identify all potentially failing paths and ensure that the design is sufficiently robust.

With a delay distribution shown in Figure 1.2b, small-delay defects pose much more of a problem than for one such as in Figure 1.2a, where the majority of paths have a large timing slack and require a relatively large-delay defect to cause failure. In Figure 1.2b, a large number of paths can be adversely affected by relatively small-delay defects.

In addition to the items discussed, the absolute magnitudes of the path delays are shrinking as clock frequencies increase. A 20-ps delay defect is

TABLE 1.1

Circuit Effects of Some Common Physical Defects

Physical Effect	Stuck at	Intermediate voltage	IDDQ	Circuit Effects				
				Delay	Cross Talk	GND Bounce	Parametric Timing Fail	VDD/Temperature Sensitive
Vt shift	no	no	no	small	small	small	yes	small
IDL variation	no	no	no	small	no	yes	yes	no
Interconnect Ω shift	no	no	no	small	small	small	yes	small
Metal xyz	no	no	no	small	small	small	yes	small
Leff shift	no	no	no	small	small	small	yes	no
Weff shift	no	no	no	small	small	small	yes	no
nMOS to pMOS length ratio	no	no	no	small	small	small	yes	no
Diffusion resistance	no	no	no	small	small	small	yes	small
Resistive via	no	no	no	small	no	no	yes	yes
Metal mouse bite	no	no	no	small	no	no	no	small
Metal silver	possible	yes	yes	small	no	no	no	yes
Line inductance	no	no	no	yes	no	yes	yes	no
Gate short–drain, –source	possible	possible	possible	small	no	no	no	no
Gate short–bulk	no	no	no	no	no	no	no	no
HCI	no	no	no	yes	no	no	no	yes

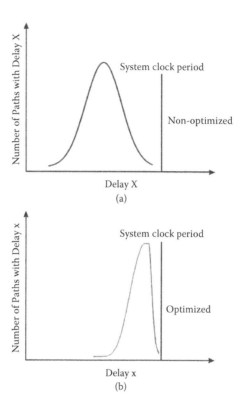

FIGURE 1.2
Distributions of path delays for (a) nonoptimized and (b) optimized designs.

unlikely to cause a problem in a 100-MHz design since it represents only 0.2% of the clock period, but for a 5-GHz design, the same defect represents 10% of the clock period and is highly likely to cause circuit failure. The optimization discussed increases the number of paths that would be sensitive to such a defect, thereby increasing the probability of failure.

1.2.5 The Interaction of Yield, Quality, and Fault Coverage

There have been a number of approaches to modeling the relationship between yield, quality, and fault coverage, with varying degrees of success. Quality can be expressed as *defect level (DL)* and defined by Equation 1.1:

$$DL = \frac{\#\ of\ defective\ parts\ which\ pass\ the\ test}{total\ \#\ parts\ which\ pass\ the\ test} \tag{1.1}$$

DL is often expressed as parts per million (ppm) or defective parts per million (DPPM). One of the earliest works [Williams 1981] deals with randomly distributed defects and uses defect coverage rather than the more traditional fault coverage. The expression derived is shown in Equation 1.2, where Y is the yield, and T is effectively the defect coverage.

$$DL = 1 - Y^{(1-T)} \qquad (1.2)$$

This was refined [de Sousa 1994] to account for defect-to-fault mapping and further refined [de Sousa 2000] to account for clustering effects. Although there are differences in the developed equations by these works and others, the trends are identical. To maintain a given quality level, if the yield decreases, the fault coverage must increase to compensate. With worst-case design, the only yield limiters were due to postdesign imperfections introduced in manufacturing, but with statistical design, some die will fail due to normal process variation. This decreases yield below what would otherwise be achievable (although from a cost perspective it still represents an improvement over the worst case because the resultant die are smaller and consume less power), which in turn places increased demands on required fault coverage.

It is important to note that fault coverage in this context is that obtained from combining a number of different types of test. Attempting to combine different tests to form a composite coverage has proven difficult [Maxwell 1994; Butler 2008], but it is clear that overall coverage is improved when a new class of defects is targeted. Because traditional delay test methods target relatively large defects, including a more thorough coverage of small-delay defects can lead to significant improvements in quality levels.

1.3 Existing Test Methods and Challenges of Smaller Geometries

The test objective is to screen defective parts from manufactured devices. Defects could modify the functionality or performance of devices. Causes for defects in manufactured devices are many. For example, open nets may be caused due to overetching or malformed vias, unwanted shorts or bridges between circuit nodes due to contaminants, and so on. For tests to detect defective parts, they should be designed such that by observing responses from a circuit under test (CUT) must lead to detection of the defects. The number of defects that could potentially occur in a manufactured device may be extremely large, and not all possible defects may be known. One attempts to use tests that can detect the defects that are most likely to occur. Methods to address detection of unknown defects have also been proposed.

Defects are physical and thus are not amenable to analysis by taking advantage of Boolean algebra used to design and analyze digital logic circuits. To facilitate analysis of defects and the derivation of tests, fault models are used to model defects. One uses the fault models to derive tests to detect the faults in the model. Even though accurate modeling of defects or their effects on the CUT is highly desirable, the most important requirement on fault models is that the tests derived to detect faults in the models detect the modeled defects. The following sections briefly review several fault models that have been used to derive tests to screen defective devices.

1.3.1 Line Stuck-at Fault Model

The line stuck-at fault model has been the most widely used fault model for more than fifty years. This fault model associates with every circuit line two faults called stuck-at-0 and stuck-at-1. When a line, say l, is stuck-at-0 (1), then independent of the circuit inputs line l will be at logic value 0 (1). Thus, a stuck-at fault may change the functionality of the circuit. Notice that a line stuck-at fault accurately models a defect that causes a zero- (low-) resistance short between a line and power supply rails. Typically, a single line in the CUT is assumed to be stuck-at-0 or stuck-at-1. In this case, the fault model is often referred to as the single (line) stuck-at fault model. If the circuit has N lines, then the number of faults in the single stuck-at fault model will have $2*N$ faults.

As an example, consider the circuit shown in Figure 1.3. This circuit has 14 lines. Note that in this model branches of a fan-out stem are considered as fault sites separate from the stem and the other branches. The single stuck-at fault model for this circuit will contain 28 faults. To reduce the number of faults that need to be explicitly targeted by test generation procedures, the fault lists are typically collapsed or reduced using equivalence relations among faults [Abramovici 1990].

A test to detect a fault must satisfy two requirements. The test must activate a fault and propagate the fault effect through the circuit to an observed output. For example, consider generating a test to detect the stuck-at-1 fault

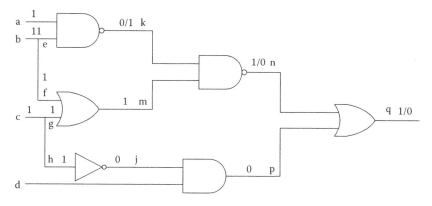

FIGURE 1.3
Test to detect stuck at 1 fault on k.

on line k in the circuit given in Figure 1.3. To activate this fault, a test should set k to 0. In the presence of the stuck-at-1 fault on k, the value on this line will be 1 instead of 0. This is indicated by the error value 0/1 representing the fault-free value of 0 and the faulty value of 1. To set k to zero, inputs a and b should be set to 1. The circuit output observed is q, and the error on k should be propagated to q. This requires propagation of the error value along the circuit path (k, n, q). This is referred to as *sensitizing* the path (k, n, q).

To sensitize a path in the circuit, the inputs to each gate along the path that are on the path need to be at what is referred to as a *noncontrolling* value or carry the same error value as the circuit line on the path. The noncontrolling value of primitive gates AND and NAND is 1, and the non-controlling value of OR and NOR is 0. Path (k, n, q) is sensitized by setting m to 1 and p to 0. This requires specifying additional circuit inputs. For example, by setting input c to 1, both m and p are set to their noncontrolling values. Thus, a test that detects k stuck-at-1 is obtained by setting $a = b = c = 1$, and input d can remain unspecified (shown as x). This test can also detect several other faults, such as a or b or e stuck-at-0, n stuck-at-0, and q stuck-at-1.

In general, for combinational circuits and full-scan circuits, tests to detect stuck-at faults are single-pattern tests. Methods to generate tests for single stuck-at faults have been studied for over fifty years and are widely used in industry as part of test suites. As the design feature sizes diminish, use of tests for additional fault models has gained importance. In the next sections, we discuss some of these fault models.

1.3.2 Bridging Fault Models

Most common defects in the VLSI devices are bridges or shorts and opens. A bridge is an unwanted (resistive) connection between two circuit nodes. Figure 1.4 illustrates a bridge/short between two outputs of G and H. R_b is the resistance of the bridge defect. The driving strengths of the gates driving the

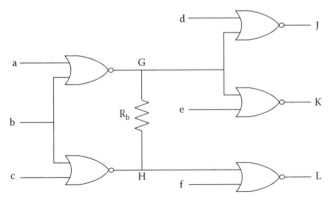

FIGURE 1.4
A bridge defect.

bridged nodes and the value of R_b determine the voltages on G and H. The driving strengths of the gates depend on the inputs applied to the gates. The gates downstream of the bridged nodes interpret the logic state of nodes G and H depending on the threshold voltages of the inputs driven by G and H. Accurate modeling of these effects using logic-level fault models is clearly complex.

Several logic-level fault models have been proposed [Abramovici 1990]. These include simpler models such as AND and OR bridging fault [Mei 1974] and four-way bridging fault models [Sengupta 1999; Krishnaswamy 2001], which are independent of the resistance of the bridge and the location of the bridged nodes. More complex models attempt to include the resistance of the bridge and the locality of the bridged nodes [Acken 1992; Maxwell 1993]. Another complication that can arise in the case of a bridge is that a feedback connection could occur. For example, if a bridge occurs between nodes a and G, the resulting feedback connection may cause an oscillation if a and b are both at 0.

As pointed out previously, detection of a fault requires activation of the fault and propagation of an error value at the fault site to an observed output. In the case of a bridge, a necessary condition to activate the fault, independent of the specific bridge fault model used, the pair of bridged nodes should be driven to opposite states. For example, to activate the bridge between nodes G and H, either G should be 1 and H 0 or G should be 0 and H 1. Additional conditions may need to be satisfied based on the specific fault model. An error occurs on either G or H or on both G and H based on the model used. In reality, if the bridge resistance is high, it may be possible that no error value can be created on either of the two bridged nodes.

As an example, consider the AND bridging fault model. In this case, the two bridged nodes take the value that is the AND of the values to which they are driven. Considering the bridge shown in Figure 1.4, if G is driven to 1 and H is driven to 0, both nodes take the value 0; hence, an error occurs on G, and the fault is detected by propagating the error on G to an observed output. Such a test is also a test to detect the stuck-at-0 fault on G. In other words, tests for bridges can be generated as tests for stuck-at faults with additional constraints satisfying the conditions to activate the modeled bridging fault.

For some values of bridge resistance, a bridge may not be detectable by tests for stuck-at faults. Instead, tests for delay faults discussed in the following sections are needed.

The previous discussion of tests for bridging faults assumes what is referred to as logic tests in which test response is measured by logic states of observed outputs. Another way to measure test responses for the detection of bridges is to measure steady-state supply currents. Such testing is called IDDQ (based) testing. In CMOS circuits, ideally the steady-state supply current IDDQ should be zero. However, when a bridge defect exists in the CUT and the bridged nodes are driven to opposite logic values, a path from the supply voltage (VDD) to ground is established through the bridge resistance and the transistors in the bridged gates. This causes nonzero IDDQ that can be used to detect bridge defects. However, with decreasing feature sizes in VLSI circuits, total leakage current is increasing, and the change in IDDQ due to a bridge defect may not be reliably detectable.

1.3.3 n-Detection

Not all likely defects in manufactured devices may be known, or they may be too numerous to explicitly target for test generation. For example, the total number of bridge defects that can occur will be large if one assumes that every pair of nodes in a circuit can be involved in a bridge defect. Even if one imposes practical proximity requirements on the pairs of nodes that can be subjected to a bridge defect, the number of bridging faults could be large. To detect unknown and untargeted defects, the n-detection test method was proposed [Pomeranz 1999]. A test set *S* for faults in a modeled fault model is called an n-detection test set if every modeled fault is detected by *n* different tests in *S*. The idea behind n-detection tests is that each different test that activates a stuck-at fault may activate a different unmodeled fault that affects the stuck-at fault site and is detected since the stuck-at fault is detected. Results from several experiments on manufactured devices have shown that n-detection tests for single stuck-at faults do help in screening out parts that passed the standard 1-detection tests. Typically, the size of the n-detection test sets grows linearly with *n*.

A method to achieve unmodeled defect coverage as n-detection test sets, called embedded multidetect tests, was also proposed [Geuzebroek 2007]. Multidetect test sets attempt to increase the number of tests that detect modeled faults while the target is detecting each fault once. The n-detection test sets for delay faults were considered [Pomeranz 1999]. Unmodeled defect coverage by n-detection test sets may not be affected when test compaction procedures are used to reduce or derive test sets with a minimal number of tests [Reddy 1996]. Probabilistic arguments for the effectiveness of n-detection test sets may need to be reassessed for future technologies, which may exhibit design- and process-dependent defects [Goel 2009].

1.3.4 Transition Fault Model

Some defects, such as resistive opens, weak transistors, and low-resistance bridges, cause signal propagation delays to increase. Process variations may also cause signal propagation delays to be higher than modeled values. Instances of increased signal propagation delays are modeled by what are called delay faults. Several delay fault models have been proposed. In this chapter and in subsequent chapters, delay faults are extensively discussed. Increased signal propagation delays may occur at one gate or over several gates. The former are modeled by *gate delay faults* and the latter by *path delay faults*. In the gate delay fault model, two delay faults, *slow to rise* (STR) and *slow to fall* (STF), are associated with each gate input, gate output, and circuit inputs. Slow-to-rise (slow-to-fall) fault models increase the delay in propagating a rising or 0-to-1 transition (falling or 1-to-0 transition) through the gates driven by the fault site.

Tests for gate delay faults require accounting for delay defect size. For example, if the defect size at a circuit lead r is less than the slack of r, the fault may not be detectable by any test. Slack of a circuit line is the difference between the period of the functional clock and the maximum delay of all paths through r. This requires accurate timing models for signal propagation and accommodation of the fact that signal propagation delays can only be modeled as a range between minimum and maximum delays [Iyengar 1988a,1988b; Pramanick 1997]. In addition, propagation delay of a gate input depends on the states of the other inputs to the gates and coupling capacitances to other adjacent lines. Another issue that crops up is the fact that more than one test may be needed to detect all defect sizes at some fault site that may cause malfunction at the desired frequency of operation [Pramanick 1997]. Methods to determine a threshold value of defect size above which the fault is detected by a given test have been developed [Iyengar 1988a,1988b; Pramanick 1997; Dumas 1993]. However, such methods pessimistically estimate the defect sizes covered [Pramanick 1997] because a given test may detect a range of defect sizes instead of only defects with sizes larger than a threshold. For gate delay faults as well as transition faults, activation of faults by hazards or glitches also needs to be considered [Pramanick 1997; Brand 1994; Pomeranz 2010].

A simplified gate delay fault model called *transition fault model or* transition delay fault (*TDF*) *model* is extensively used. In this fault model, the delay at the fault site is assumed to be large, and the signal propagation delay of every circuit path through the fault site exceeds the clock period. This makes test generation and determination of fault coverage simpler. Tests for TDFs can be generated by straightforward modification of procedures for generating tests to detect line stuck-at faults [Waicukauski 1987]. For example, consider detection of a STR fault on line d in the circuit shown in Figure 1.5. To activate this fault, we should create a rising transition on line d. This implies that the fault can only be activated by applying two consecutive patterns. A pair of such patterns is shown in Figure 1.5a. In the presence of an STR fault on line d, the value on d will not change to 1 from 0 before the circuit outputs are read due to the assumption that the delay defect size is large enough. Thus, when the circuit output is read, line d will still be 0. This is shown as the faulty value under the slash in Figure 1.5b. Thus, the STR TDF manifests itself as a stuck-at-0 fault when the second input of the two-pattern test is applied. In general, to detect an STR(STF) TDF a two-pattern test <t1, t2> must satisfy the following two conditions: (1) the first pattern t1 should set the value on the faulty line to 0(1), and (2) the second pattern t2 must detect a stuck-at-1(0) fault on the faulty line. The first pattern t1 is called the initialization pattern that initializes the faulty line to a 0(1) for THE STR(STF) TDF fault. Thus, one can modify test pattern generators and fault simulators for stuck-at faults in a

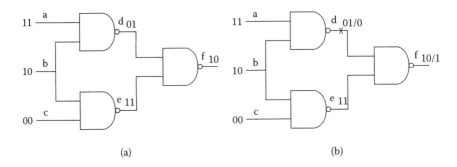

FIGURE 1.5
Detection of a STR TDF.

straightforward manner to obtain similar tools for TDFs. This is one of the advantages of the TDF model.

1.3.5 Path Delay Fault Model

The path delay fault model [Smith 1985] is more comprehensive since it can accommodate spot defects and distributed defects. A path delay fault is associated with each logical path and is said to be present if the delay of the logical path exceeds the slack of the path. Tests to detect path delay faults are classified according to the conditions satisfied by the side inputs of the gates in the path. The different types of tests are discussed next.

In Figure 1.6, a robust test for the logical path *c-g-h-k* with rising transition at its input is shown. The path delay fault test is called a robust test if it detects the fault independent of the delays in the rest of the circuit [Smith 1985; Lin 1987]. Figure 1.6 shows the signal values for the two-pattern test where S0 represents a signal value that is a glitch-free 0 during the application of the two patterns. For the same fault, in Figures 1.6 and 1.7 nonrobust tests called strong nonrobust and weak nonrobust tests are shown. The signal value H0 represents a signal value that is a 0 in the steady state but may have a hazard or glitch during the transition from the first pattern to the

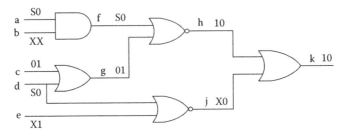

FIGURE 1.6
Robust test for path *c-g-h-k* with rising transition.

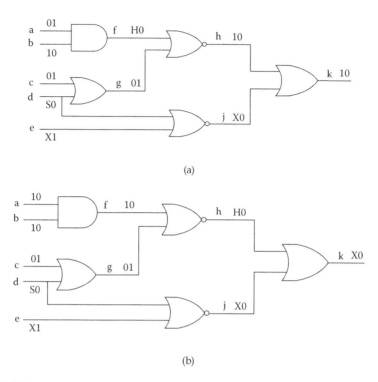

(a)

(b)

FIGURE 1.7
(a) Strong nonrobust test for path *c-g-h-k* with rising transition. (b) Weak nonrobust test for path *c-g-h-k* with rising transition.

second pattern of the two-pattern test. Nonrobust tests can be invalidated due to circuit delays that affect signals at the off-path inputs [Konuk 2000; Pomeranz 2008]. For some paths, neither robust nor nonrobust tests may exist. However, such paths may be functional paths and should be tested using functional sensitization [Cheng 1996].

Generation of tests for path delay faults and fault simulation have been extensively investigated [Krstic 1998; Pomeranz 1998; Bushnell 2000; Jha 2003]. Typically, one uses the necessary conditions on off-path inputs of the gates on the path for sensitizing the path to determine additional necessary conditions through implications followed by justifying all the necessary conditions. Path sensitization conditions for different types of two-pattern tests are shown in Table 1.2. In this table, S0 (S1) represent hazard-free 0(1) during the application of the two-pattern test, and H0 (H1) represent a signal that is a 0 (1) in the steady state but may have a hazard/glitch.

1.3.6 Test Implementation and Adaptive Test

Process variation was discussed in Section 1.2.2; it was noted that circuit behavior can change within a wide margin. Further, these changes can exist

TABLE 1.2

Sensitization Conditions on Off-Path Inputs

	Robust		Nonrobust Strong		Weak		Functional	
	AND	OR	AND	OR	AND	OR	AND	OR
Transition	NAND	NOR	NAND	NOR	NAND	NOR	NAND	NOR
0 → 1	X1	S0	X1	H0	X1	X0	X1	XX
1 → 0	S1	X0	H1	X0	X1	X0	XX	X0

within a die, from die to die, and from wafer to wafer. The delay test methods already discussed in themselves do not take such variation into account, and it is the method of implementing them that must meet the challenges imposed. Since design cannot control for all variation and uncertainty, test methods must identify variation and screen for it. Traditional tests cannot accomplish this since they rely on setting a static test limit when the parameter under measurement varies widely for parts belonging to the intrinsic population. Setting the test limit to the worst case results in missing many defective die. It is essential that limits are able to be adapted to the DUT to improve defect detection.

The concept of using device response to adapt a test in this way became known as data-driven testing and formed the basis of a great deal of research into outlier screening methods [Daasch 2005]. Considering the delay test, a large defect is likely to produce a failure regardless of path length variation; a small-delay defect might produce a failure on a slow die but not on a fast die. However, on this fast die the existence of the small defect could result in significantly reduced timing margins, causing potential failures under different operating conditions. The detection of the defect is therefore important. Some methods of doing so are discussed further in this book. Outlier detection approaches are also possible; one example is to measure an on-chip ring oscillator frequency and use that to predict F_{max} (maximum operating frequency). A defective chip is likely to have a lower F_{max} than predicted and lie outside the intrinsic population.

1.4 Effect of Small Delays on Transition Testing

The transition fault model allows simpler test generation and fault simulation procedures. However, since the delay defect size is not explicitly considered, a test for a transition fault may propagate the fault effect through a short path with large slack, and the associated delay fault may not be detected if

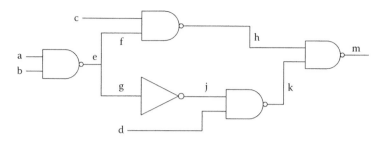

FIGURE 1.8
Detection of small-delay defect.

its size is less than the slack. For example, consider detection of a delay fault on line e in the circuit shown in Figure 1.8. The fault effect can be propagated through path f-h-m or through path g-j-k-m. Since the delay of path g-j-k-m is larger, a smaller defect can be detected compared to the defect size that can be detected if the fault effect is propagated through path f-h-m.

The example in Figure 1.8 illustrates the need to generate tests that propagate fault effects along paths that pass through the fault site and have the largest delay. This way, the smallest detectable delay defects at the fault sites can be detected. One way to ensure detection of all detectable delay defects that affect correct circuit operation at a desired clock frequency is to test all path delay faults. However, such an approach is not practical since the number of paths in circuits are typically extremely large. In addition, most paths may not be robustly testable. An approach related to use of path delay faults is to select a subset of paths to test. Criteria to select paths have been investigated and are discussed in Chapter 2. Recent works to address tests of small-delay defects have focused on enhancing transition fault tests. Generation of such tests requires the automatic test pattern generation (ATPG) tool to be aware of the circuit delays. Such an ATPG is called a *timing-aware ATPG,* and the tests generated by a timing-aware ATPG are called *timing-aware tests.* A timing-aware test is discussed in detail in Chapter 3. Test methods based on a faster-than-at-speed approach are described in Chapters 4 and 5. In the faster-than-at-speed testing, existing or modified transition fault patterns are applied at a clock speed higher than the system clock. The application of a test pattern at a higher clock speed reduces the available system slack on all paths, thereby increasing the chances of detection of small-delay defects on all paths. Alternative or hybrid test methods that resolve the issue of the large pattern count and run time without compromising the test quality are described in Chapters 6, 7, and 8. Most of the alternative test methods are based on either effective test pattern selection or effective fault site selection. Finally, a detailed analysis of different test quality metrics used to measure the small-delay defect effectiveness for a given pattern set is presented in Chapter 9.

References

[Abramovici 1990] M. Abramovici, M.A. Breuer, and A.D. Friedman, *Digital Systems Testing and Testable Design*, Wiley-IEEE Press, New York, 1990.

[Acken 1992] J.M. Acken and S.D. Millman, Fault model evolution for diagnosis accuracy versus precision, in *Proceedings Custom Integrated Circuits Conference*, 1992, pp. 13.4.1–13.4.4.

[Brand 1994] D. Brand and V.S Iyengar, Identification of redundant delay faults, *IEEE Transactions on CAD of Integrated Circuits and Systems*, 13(5), pp. 553–565, 1994.

[Braun 2005] A.E. Braun, Pattern-related defects become subtler deadlier, *Semiconductor International*, 1 June 2005.

[Bushnell 2000] M.L. Bushnell and V.D. Agrawal, *Essentials of Electronic Testing for Digital, Memory and Mixed-Signal VLSI Circuits*, Springer, New York, 2000.

[Butler 2008] K.M. Butler, J.M Carulli, and J. Saxena, Modeling test escape rate as a function of multiple coverages, in *Proceedings International Test Conference*, 2008.

[Cheng 1996] K.-T. Cheng and H.-C. Chen, Classification and identification of non-robustly untestable path delay faults, *IEEE Transactions on Computer-Aided Design Integrated Circuits and Systems*, 15(8), pp. 845–853, August 1996.

[Daasch 2005] W.R. Daasch and R. Madge, Data-driven models for statistical testing: measurements, estimates and residuals, in *Proceedings IEEE International Conference*, 2005, p. 10.

[de Sousa 1994] J.T. de Sousa, F.M. Goncalves, J.P. Teixeira, and T.W. Williams, Fault modeling and defect level projections in digital ICs, in *Proceedings European Design and Test Conference (EDTC)*, 1994, pp. 436–442.

[de Sousa 2000] J.T. de Sousa and V.D. Agrawal, Reducing the complexity of defect level modeling using the clustering effect, in *Proceedings Design, Automation and Test in Europe Conference and Exhibition*, 2000, pp. 640–644.

[Dumas 1993] D. Dumas, P. Girard, C. Landrault, and S. Pravossoudovitch, An implicit delay fault simulation method with approximate detection threshold calculation, in *Proceedings International Test Conference*, 1993, pp. 705–713.

[Geuzebroek 2007] J. Geuzebroek, E.J. Marinissen, A.K. Majhi, A. Glowatz, and F. Hapke, Embedded multi-detect ATPG and its effect on the detection of unmodeled defects, in *Proceedings International Test Conference*, 2007, pp. 1–10.

[Goel 2009] S.K. Goel, N. Devta-Prasanna, and M. Ward, Comparing the effectiveness of deterministic bridge fault and multiple-detect stuck fault patterns for physical bridge defects: A simulation and silicon study, in *Proceedings International Test Conference*, 2009, pp. 1–10.

[Hess 2010] C. Hess, A. Inani, A. Joag, Zhao Sa, M. Spinelli, M. Zaragoza, Nguyen Long, and B. Kumar, Stackable short floe characterization vehicle test chip to reduce test chip designs, mask cost and engineering wafers, in *Proceedings Advanced Semiconductor Manufacturing Conference*, 2010.

[ITRS 2007] International Technology Roadmap for Semiconductor Industry home page, 2007, http://www.itrs.net/Links/2007ITRS/Home2007.htm

[Iyengar 1988a] V.S. Iyengar, B.K. Rosen, and I. Spillinger, Delay test generation I—Concepts and coverage metrics, in *Proceedings IEEE International Test Conference*, 1988, pp. 857–866.

[Iyengar 1988b] V.S. Iyengar, B.K. Rosen, and I. Spillinger, Delay test generation II—Concepts and coverage metrics, in *Proceedings IEEE International Test Conference*, 1988, pp. 867–876.

[Jha 2003] N.K. Jha and S.K. Gupta, *Testing of Digital Systems*, Cambridge University Press, Cambridge, UK, 2003.

[Killpack 2008] K. Killpack, S. Natarajan, A. Krishnamachary, and P. Bastani, Case study on speed failure causes in a microprocessor, *IEEE Design and Test of Computers*, 25(3), pp. 224–230, May–June 2008.

[Konuk 2000] H. Konuk, On invalidation mechanisms for non-robust delay tests, in *Proceedings International Test Conference*, 2000, pp. 393–399.

[Krishnaswamy 2001] V. Krishnaswamy, A.B. Ma, and P. Vishakantaiah, A study of bridging defect probabilities on a Pentium™ 4 CPU, in *Proceedings International Test Conference*, 2001, pp. 688–695.

[Krstic 1998] A. Krstic and K.-T. Cheng, *Delay Fault Testing for VLSI Circuits*, Springer, Boston, 1998.

[Kuhn 2008] K. Kuhn, C. Kenyon, A. Kornfield, M. Liu, A. Maheshwari, W.-K. Shih, S. Sivakumar, G. Taylor, P. VanDerVoorn, and K. Zawadzki, Managing process variation in Intel's 45nm CMOS technology, *Intel Technology Journal*, 2(17), 2008.

[Kulkarni 2006] S.H. Kulkarni, D. Sylvester, and D. Blaauw, A statistical framework for post-silicon tuning through body bias clustering, in *Proceedings of IEEE/ACM International Conference on Computer-Aided Design (ICCAD)*, 2006, pp. 39–46.

[Lin 1987] C.J. Lin and S.M. Reddy, On delay fault testing in logic circuits, *IEEE Transactions on Computer-Aided Design*, 6(9), pp. 694–701, September 1987.

[Mahmoodi 2005] H. Mahmoodi, S. Mukhopadhyay, and K. Roy, Estimation of delay variations due to random-dopant fluctuations in nanoscale CMOS circuits, *IEEE Journal of Solid-State Circuits*, 40(9), pp. 1787–1796, September 2005.

[Maxwell 1993] P.C. Maxwell and R.C. Aitken, Biased voting: A method for simulating CMOS bridging faults in the presence of variable gate logic thresholds, in *Proceedings International Test Conference*, 1993, pp. 63–72.

[Maxwell 1994] P.C. Maxwell, R.C. Aitken, and L.M. Huisman, The effect on quality of non-uniform fault coverage and fault probability, *ITC1994*, pp. 739–746.

[Mei 1974] K. Mei, Bridging and stuck-at faults, *IEEE Transactions on Computers*, C-23(7), pp. 720–727, 1974.

[Meijer 2012] M. Meijer, and J.P. Gyvez, Body-bias-driven design strategy for area- and performance-efficient CMOS circuits, *IEEE Transactions on VLSI Systems* 20(1), pp. 42–51, 2012.

[Park 1991] E.S. Park, B. Underwood, T.W. Williams, and M. Mercer, Delay testing quality in timing-optimized designs, in *Proceedings International Test Conference*, October 1991, pp. 897–905.

[Pomeranz 1998] I. Pomeranz and S.M. Reddy, A generalized test generation procedure for path delay faults, in *Proceedings International Symposium on Fault-Tolerant Computing*, 1998, pp. 274–283.

[Pomeranz 1999] I. Pomeranz and S.M. Reddy, On n-detection test sets and variable n-detection test sets for transition faults, in *Proceedings IEEE VLSI Test Symposium*, 1999, pp. 173–180.

[Pomeranz 2008] I. Pomeranz and S.M. Reddy, Transition path delay faults: A new path delay fault model for small and large delay defects, *IEEE Transactions on Very Large Scale Integration (VLSI) Systems*, 16(1), pp. 98–107, January 2008.

[Pomeranz 2010] I. Pomeranz and S.M. Reddy, Hazard-based detection conditions for improved transition path delay fault coverage, *IEEE Transactions on CAD of Integrated Circuits and Systems*, 18(2), pp. 332–337, 2010.

[Pramanick 1997] A.K. Pramanick and S.M. Reddy, On the fault detection coverage of gate delay fault detecting tests, *IEEE Transactions on Computer-Aided Design*, 16(1), pp. 78–94, 1997.

[Raina 2006] R. Raina, What is DFM and DFY and why should I care? In *Proceedings International Test Conference*, 2006, pp. 1–9.

[Reddy 1996] S.M. Reddy, I. Pomeranz, and S. Kajihara, On the effects of test compaction on defect coverage, in *Proceedings VLSI Test Symposium*, 1996.

[Segura 2002] J. Segura, A. Keshavarzi, J.M. Soden, and C.F. Hawkins, Parametric failures in CMOS ICs—A defect-based analysis, in *Proceedings International Test Conference*, 2002, pp. 90–99.

[Sengupta 1999] S. Sengupta et al., Defect-based tests: A key enabler for successful migration to structural test, *Intel Technology Journal*, Quarter 1, 1999.

[Smith 1985] G.L. Smith, Model for delay faults based upon paths, in *Proceedings IEEE International Test Conference*, 1985, pp. 342–349.

[Smith 1997] M. Smith, *Application-Specific Integrated Circuits*, Addison-Wesley Professional, Reading, MA, 1997.

[Srivastava 2005] A. Srivastava, S. Shah, K. Agrawal, D. Sylvester, D. Blaauw, and S. Director, Accurate and efficient gate-level parametric yield estimation considering correlated variations in leakage power and performance, in *Proceedings of Design Automation Conference*, 2005.

[Waicukauski 1987] J.A. Waicukauski, E. Lindbloom, B.K. Rosen, and V.S. Iyengar, Transition fault simulation, *IEEE Design and Test of Computer*, pp. 32–38, April 1987.

[Williams 1981] T.W. Williams and N.C. Brown, Defect level as a function of fault coverage, *IEEE Transactions on Computers*, C-30, pp. 987–988, 1981.

[Williams 1991] T.W. Williams, B. Underwood, and M.R. Mercer, The interdependence between delay-optimization of synthesized networks and testing, in *Proceedings ACM/IEEE Design Automation Conference*, 1991, pp. 87–92.

[Xie 2009] Y. Xie and Y. Chen, Statistical high level synthesis considering process variations, *IEEE Computer Design and Test*, Special Issue on HLS, 26(4), pp. 78–87, July–August, 2009.

Section I

Timing-Aware ATPG

2

K *Longest Paths*

Duncan M. (Hank) Walker

CONTENTS

2.1 Introduction

Delay testing detects small manufacturing defects that do not cause functional failure but affect the speed of integrated circuits. The path delay fault model [Smith 1985] is the most conservative of the classical delay fault models because a circuit is faulty if any path delay exceeds the specification time. The problem with this model is that the paths in real circuits cannot be enumerated. To overcome this problem, some test methods only cover a subset of the paths (e.g., the global longest paths) [Lin 1987; Bell 1996] or the longest path through each gate [Li 1989; Majhi 2000; Murakami 2000; Shao 2002; Sharma 2002]. A delay fault caused by a local defect, such as a resistive open or short, can only be detected by testing a path through it, and testing the longest path through it can detect the smallest local delay defect. Process variation and noise can result in the longest path through a gate varying from chip to chip or across operating conditions. Therefore, testing only one path through each gate cannot guarantee the detection of the smallest local

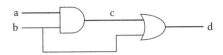

FIGURE 2.1
A circuit with a false path *a-c-d*.

delay faults. Testing the *K* longest paths through a fault site increases the fault detection probability.

A path is said to be *testable* if a transition can propagate from the primary input (PI) to the primary output (PO) associated with the path under certain sensitization criteria [Smith 1985; Lin 1987; Benkoski 1990; Chang 1993]. If a path is not testable, it is called an *untestable* or *false* path. For example, in Figure 2.1, path *a-c-d* is a false path under the single-path sensitization criterion [Benkoski 1990] because to propagate a transition through the AND gate requires line *b* to be logic 1, and to propagate the transition through the OR gate requires line *b* to be logic 0.

There has been significant prior work on global longest-path generation, but this work has not been applied to the problem of generating the *K* longest paths per gate (KLPG). NEST [Pomeranz 1995] generates paths in a non-enumerative way but is only effective in highly testable circuits. DYNAMITE [Fuchs 1991] is efficient in poorly testable circuits but has high memory consumption in highly testable circuits. RESIST [Fuchs 1994] exploits the fact that many paths in a circuit have common subpaths and sensitizes those subpaths only once, which reduces repeated work and identifies large sets of untestable paths. Moreover, for the first time this research identified 99.4% of all the path delay faults as either testable or untestable in circuit c6288, which is known for having an exponential number of paths. However, test generation for c6288 took 1,122 CPU (central processing unit) hours to find 12,592 paths on a SPARC IPX 28-MIPS (million-instructions-per-second) machine. The RESIST algorithm was extended to the problem of finding a set of longest testable paths that cover every gate [Sharma 2002]. This method takes advantage of the relations between the longest paths through different gates and guarantees their testability. However, this work assumes a unit delay model; it does not handle circuit c6288, and there is no clear way to extend the algorithm to generate the KLPG. An as-late-as-possible transition fault (ALAPTF) approach was proposed to launch one or more transitions at the fault site as late as possible to detect faults through the least-slack paths [Gupta 2004].

A statistical static timing analysis tool [Bell 1996] presents another method to efficiently identify the global longest testable paths in a combinational circuit. Instead of generating many long structural paths and checking their testability, this algorithm grows paths from the PIs. Each iteration adds a new gate and applies side input constraints. Then, instead of assigning logic values on one or more PIs to satisfy the constraints on the newly added gate, as done in VIPER [Chang 1993], direct implications are applied to find local conflicts. If conflicts exist, the whole search space that contains the already-grown partial path is trimmed. This technique is called implicit false path elimination [Benkoski 1990; Stewart 1991]. Some other false path elimination techniques, such as forward trimming and dynamic dominators, are also applied to identify false paths earlier in the search process. This tool is efficient and able to generate globally longest paths for circuit c6288.

This algorithm was extended to generate the KLPG test set for a combinational circuit. It inherits the framework of Bell [1996] but aims at particular gates one by one. This algorithm also takes advantage of the relation between the long paths through different gates, which was revealed in previous research [Sharma 2002] and extended in this work, to reduce the search space and avoid repeated work. In this work, we also determined that initially applying the global longest path generation could cover some gates quickly.

2.2 Path Generation for Combinational Circuits

Prior to path generation, topological circuit information is collected to help guide the path generation process and trim the search space. First, the min-max delay from each gate to a PO is computed without considering any logic constraint (termed the *PERT delay*). A gate's min-max PERT delay can be simply computed using its fan-out gates' min-max PERT delays and the rising/falling buffer-to-buffer delays between gates.

In addition to the PERT delays, the earliest and latest possible rising/falling transition times on the input and output lines for each gate are computed, assuming that a transition at any PI can only occur at time zero. This procedure is similar to the PERT delay computation, with complexity linear in the circuit size. This information is useful under some sensitization criteria because transitions can occur only within the earliest/latest range. When propagating a transition that transits to a controlling value through a gate, if a side input cannot have a transition before the on-path transition, the final value on that side input can be either a controlling or a noncontrolling value (the initial value must still be noncontrolling), without blocking the on-path transition. Without this information, the search procedure may unnecessarily require the final value on the side input to be noncontrolling.

To find the K longest testable paths through gate g_i, a path store is established for the path generation. In the path store, many *partial paths*, which may become the K longest testable paths through gate g_i, are stored. A partial path is a path that originates at a PI but has not reached a PO. Figure 2.2 shows an example. The partial path starts from PI g_0, and ends at gate g_i. Initially, the path store contains $2n_{PI}$ partial paths, where n_{PI} is the number of PIs. There are two partial paths from each PI, representing a rising or

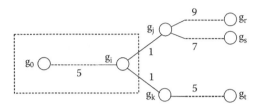

FIGURE 2.2
A partial path and its esperance.

falling transition at that PI. Each partial path initially has only one node (a PI). A partial path grows when a gate is added to it. When a partial path reaches a PO, it becomes a *complete path*.

A value called *esperance* [Benkoski 1990] is associated with a partial path. The min-max esperance is the sum of the length of the partial path and the min-max PERT delay from its last node to a PO. In other words, the max esperance of a partial path is the upper bound of its delay when it grows to a complete path, and the min esperance is the lower bound if the complete path is testable. In Figure 2.2, suppose the length of the partial path $g_0 \dots g_i$ is 5, and the PERT delays between g_i and POs g_r, g_s, and g_t are 10, 8, and 6, respectively. The min-max esperance of partial path $g_0 \dots g_i$ is 11/15.

The path generator selects the partial path with the largest max esperance. Potentially, this partial path will grow to a complete path with maximum delay.

Figure 2.3 is the algorithm of finding the K longest testable paths through gate g_i. Before the path generation for gate g_i, all gates that are not in g_i's fan-in or fan-out cone are identified because when a partial path grows, it is impossible for any of these gates to be added to the path (otherwise, the partial path has no chance to pass through g_i). But, these gates can still determine side input constraints of a gate on the path.

Each iteration of the path generation begins by popping the partial path with max esperance from the path store. The partial path is extended by adding a fan-out gate that contributes to the largest max esperance. For example, in Figure 2.4, the partial path $g_0 \dots g_i$ is extended by adding gate g_j because extending to g_j maintains the max esperance. If the partial path has more than one extendable fan-out, it must be saved in another copy, and in the copy, the already-tried fan-out must be marked "blocked" or "tried." Then, the copy gets its esperance updated and is pushed into the path store. For example (Figure 2.4), since gate g_i has two fan-outs, and extending the partial path to g_j may later result in false paths, the partial path $g_0 \dots g_i$ must be saved because extending it to gate g_k may find a longer testable path. And, because fan-out g_j has been tried, in the copy the min-max esperance becomes 11/11.

After the partial path is extended ($g_0 \dots g_i g_j$ in Figure 2.4), the constraints to propagate the transition on the added gate (g_j) are applied. Under the *nonrobust* sensitization criterion, noncontrolling final values on the side inputs are required. Under the *robust* sensitization criterion, in addition to

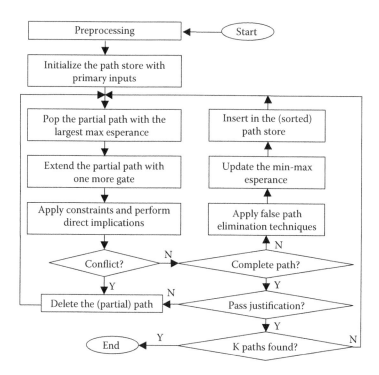

FIGURE 2.3
Path generation algorithm.

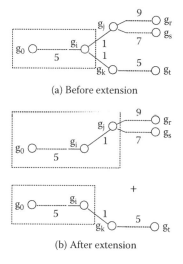

FIGURE 2.4
Extending a partial path.

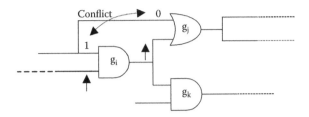

FIGURE 2.5
Conflict after applying direct implications.

noncontrolling final values, the side inputs must remain noncontrolling if the on-path input has a transition to the controlling value. Then, direct implications are used to propagate the constraints throughout the circuit. A direct implication on a gate is one where an input or output of that gate can be directly determined from the other values assigned to that gate. If a conflict happens during direct implications, the partial path is false. In other words, any path including this partial path is a false path. For example (Figure 2.4), if extending partial path $g_0 \dots g_i$ to gate g_j results in a conflict (Figure 2.5 shows an example), both path $g_0 \dots g_r$ and $g_0 \dots g_s$ are determined to be false. Therefore, the partial path is deleted from the path store so that the whole search space that contains this partial path is trimmed. Previous research [Benkoski 1990] showed that most false paths can be eliminated by direct implications.

If the extended partial path reaches a PO, it becomes a *complete path*. In this case, final justification is performed on the path using a PODEM- [Goel 1981] style algorithm. One reason to do final justification is to find a vector pair that sensitizes this path. The other reason is that some false paths do not have any conflict in the direct implications during path growth. Figure 2.6 shows an example [Benkoski 1990]. Suppose both AND gates need their side inputs to be logic 1, and direct implications stop at the two OR gates. This path does not fail the direct implications, but it is a false path.

If the extended partial path is not a complete path, some false path elimination techniques, which are discussed in detail further in this chapter, are applied to it to more efficiently prevent the new partial path from becoming a false path. Then, the min-max esperance of the partial path is updated, and it is inserted into the path store. Since its max esperance may decrease and min esperance may increase after extension, in the next iteration, another partial

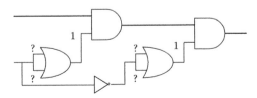

FIGURE 2.6
A path that passes direct implications but fails final justification.

path may be selected for extension. For example (Figure 2.4), after extending partial path $g_0 \ldots g_i$ to gate g_j, the min-max esperance changes from 11/15 to 13/15. If path $g_j \ldots g_r$ is blocked, which means path $g_0 \ldots g_r$ is a false path, after applying the false path elimination techniques, the min-max esperance is 13/13.

Because each partial path consumes memory, and the path store cannot have an infinite size, when the number of partial paths exceeds the path store size limit, some partial paths with low max esperance are removed from the path store. The maximum esperance of the removed partial paths is recorded. If a path is later found whose length is less than the maximum esperance, then it may not be one of the K longest testable paths. In practice, the path store can hold enough partial paths that this situation is rare.

The path generation iteration stops when K longest testable paths through gate g_i are found or the path store is empty. Since the K longest testable paths through different gates may overlap, every time a new path is generated, it must be checked to see if it has already been generated during the path generation for another gate.

2.2.1 Refined Implicit False Path Elimination

When a partial path grows, in some cases after the constraints are applied and the direct implications are performed, the possibility to continue extending the partial path to the POs may be reduced. In the extreme case, there is no way to continue extending the partial path (completely blocked).

Figure 2.7 shows an example in which the path through the logic block cannot propagate to the output. The partial path has grown to the NAND gate g_i, and the side input must be a logic 1 if the single-path sensitization constraints are considered. This value propagates forward through the inverter and becomes a controlling value on an input of NAND gate g_j. This blocks propagation from g_i to g_j through any paths within the logic block. With *forward trimming*, the entire logic block is trimmed, and the search process is guided toward an unblocked path (the upper inverter) without attempting to traverse through all the paths in the logic block.

During circuit preprocessing, the PERT delays are computed assuming no path is blocked. With the growth of a partial path, more and more

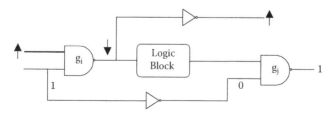

FIGURE 2.7
Application of forward trimming.

information is known because more constraints are applied. Forward trimming recomputes the min-max PERT delay from the end of the partial path based on the circuit structure and the current value assignments. In this case, blocked extension choices are not considered in the computation, and false paths can be eliminated earlier, so that the search can be guided more accurately toward a testable path. If the search space is partially trimmed, the partial path still has a chance to become a testable complete path, but its esperance may be reduced according to its PERT delay reduction. In the next iteration of path generation, a more promising partial path may be selected.

If the PERT delays are used, a local conflict in the unexplored search space is not detected until the partial path grows to that site because the PERT delays are computed without considering any logic constraint. We have developed a heuristic to exclude untestable subpaths due to local conflicts when computing the PERT delay for a gate. We call the new values *Smart-PERT* delays or *S-PERT*. For simplicity, only maximum PERT and S-PERT delays are discussed. Because some untestable subpaths are not included in the S-PERT computation, a gate's S-PERT delay is always less than or equal to its PERT delay. Moreover, compared to the PERT delay, the S-PERT delay is closer to the delay of the longest testable path from that gate to a PO.

A gate's PERT delay can be computed using its fan-out gates' PERT delays. If the unit delay model is used, $\text{PERT}(g_i) = \max \{\text{PERT}(g_j) \mid g_j$ is a fan-out gate of $g_i\} + 1$. Figure 2.8a shows an example, assuming $\text{PERT}(g_3) = 8$ and $\text{PERT}(g_4) = 6$ are known. In this example, $\text{PERT}(g_0) = 10$ is computed using $\text{PERT}(g_1)$ and $\text{PERT}(g_2)$.

When S-PERT(g_i) is computed, a user-defined variable *S-PERT depth* is used. If the S-PERT depth is set to d, then S-PERT(g_i) is computed using S-PERT(g_j), where g_j is d gates from g_i in g_i's fan-out tree. For example, in Figure 2.8b if d is set to 2, then S-PERT(g_0) is computed using S-PERT(g_3) and S-PERT(g_4).

The heuristic works as follows. Suppose S-PERT(g_i) is being computed. $G = \{g_j \mid g_j$ is d gates from g_i in g_i's fan-out tree$\}$, and G is sorted by S-PERT(g_j) in decreasing order. The heuristic pops the first gate g_j in G and attempts to propagate a transition from g_i to g_j. If there is no conflict (the transition

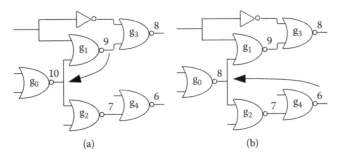

(a)   (b)

FIGURE 2.8
Computation of PERT delay (a) and S-PERT delay (b).

successfully reaches g_j, with all the constraints applied), S-PERT(g_i) is set to S-PERT(g_j) + d. Otherwise, it pops the second gate in G and repeats the same procedure. In Figure 2.8b, for example, at first the heuristic tries to propagate a transition from g_0 to g_3 but finds it is impossible to set the side inputs of g_1 and g_3 both to noncontrolling values. Then, it tries g_4 and does not meet any conflict. So, S-PERT(g_0) is 8. It is obvious that increasing the S-PERT depth can make the S-PERT delays closer to the delay of the longest testable path from that gate to a PO, but its cost increases exponentially. In this work, the S-PERT depth is fixed, but one option is to increment it if path searches repeatedly fail.

The usefulness of S-PERT is highly dependent on the circuit structure. The most benefit would be derived from a path with d gates, each of which has fan-out f, and each fan-out reconverges at a later gate (Figure 11). This results in f^d possible paths that must be traversed. The worst case is when all of them are false but conflicts do not occur until the partial path grows very long. With the use of S-PERT delays, the path generation extends to a shorter structural path P_2 (Figure 2.9) because it has larger esperance, and the traversal of all the false paths with equal length from g_1 to g_4 is avoided. As our experimental results show, this technique helps exclude many false paths in circuit c6288.

There are close relations between long testable paths through different gates because a long testable path through one gate is likely a long testable path through another gate [Sharma 2002]. This can be exploited during path generation. We allocate two arrays for each gate in the circuit: $L_{ub}[1 \ldots K]$ and $L_{lb}[1 \ldots K]$, which indicate the upper and lower bound, respectively, of the lengths of the K longest testable paths through this gate. The two arrays are sorted. Initially, the values in $L_{ub}[1 \ldots K]$ are all set to the length of the longest structural path through the gate, and the values in $L_{lb}[1 \ldots K]$ are all 0.

When the K longest testable paths are found for gate g_i, $L_{ub}[1 \ldots K]$ and $L_{lb}[1 \ldots K]$ for g_i are updated to the actual lengths of the K longest testable paths. Suppose a newly found path for gate g_i also contains gate g_j, which means this path passes through both g_i and g_j. If the length of this path is greater than that of a previously found path for g_j, $L_{lb}[1 \ldots K]$ for g_j is updated by inserting a link to the newly found path and deleting the link to the

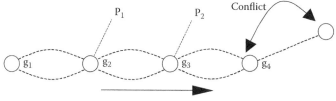

Path extension direction

FIGURE 2.9
A circuit with an exponential number of false paths.

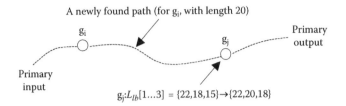

FIGURE 2.10
Updating $L_{lb}[1 \dots K]$.

shortest path found. This process may increase the values in $L_{lb}[1 \dots K]$ for all the gates contained in the newly found path. Figure 2.10 shows an example. Assuming $K = 3$, at some point $L_{lb}[1 \dots 3]$ for gate g_j is {22, 18, 15}, which means the lengths of the three longest paths through g_j found during path generation are at least 22, 18, and 15. Suppose the length of a newly found path for gate g_i is 20. Then, $L_{lb}[1 \dots 3]$ for gate g_j is updated to {22, 20, 18}.

On the other hand, the values in $L_{ub}[1 \dots K]$ for some gates decrease when a new path is found. Suppose gate g_i has f fan-in gates, and $\cup_{fanin}L_{ub}[1 \dots K]$ indicates the union of the $L_{ub}[1 \dots K]$ arrays of its fan-in gates, and it is sorted in decreasing order. The upper bound of the lengths of the K longest testable paths through gate g_i cannot exceed the first K values in $\cup_{fanin}L_{ub}[1 \dots K]$ because all the paths through g_i must also pass through one of its fan-in gates. Figure 2.11 shows an example. Assuming $K = 3$ and gate g_i has two fan-in gates, with $L_{ub}[1 \dots 3]$ = {17, 16, 11} and {20, 18, 12}. Then, the values in $L_{ub}[1 \dots 3]$ for g_i must be no more than {20, 18, 17}. The same analysis can be performed using the fan-out gates or absolute dominators [Bell 1996] of gate gAs more paths are found, the values in $L_{ub}[1 \dots K]$ and $L_{lb}[1 \dots K]$ for gate g_i, for which the path generation has not been performed, become closer. If they are close enough, say less than 1% difference, it can be assumed that the K longest testable paths for gate g_i have been found, so the path generation for

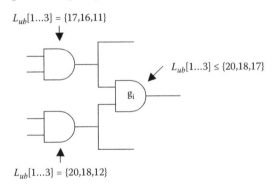

FIGURE 2.11
Updating $L_{ub}[1 \dots K]$.

it can be skipped. Many gates can be skipped if a unit delay model is used [Sharma 2002].

If gate g_i cannot avoid path generation, during its path generation process, if the max esperance of the partial path being processed is less than its $L_{lb}[v]$ ($1 < v \leq K$) value, then the first v paths already found for g_i are proven to be the v longest paths through g_i because the partial paths in the path store have no chance to grow to a complete path with larger length. So, $L_{ub}[u]$ ($u = 1, \ldots, v$) are updated accordingly (set to $L_{lb}[u]$). On the other hand, when the vth longest path through gate g_i is being searched, and the min esperance of a partial path is greater than $L_{ub}[v]$, the partial path can be deleted immediately because when it becomes a complete path, this path is either a false path or already found.

The $L_{ub}[1]$ values of other gates can also be taken advantage of during path generation for gate g_i. Suppose a partial path grows to gate g_j. If the max esperance of the partial path is greater than $L_{ub}[1]$ for g_j, it must be reduced to $L_{ub}[1]$ for g_j because it is impossible for this partial path to grow to a complete path with length greater than $L_{ub}[1]$ for g_j; otherwise, this path would also be a testable path through g_j, which invalidates $L_{ub}[1]$ for g_j. When the partial path continues to grow, its min esperance may increase, and if it becomes larger than its max esperance, the partial path is deleted because it must eventually grow to a false path.

The *global longest paths* are the longest paths throughout the circuit, regardless of which gates they pass through. The global longest path generation algorithm is a slight modification to the path generation algorithm for a particular gate (Figure 2.3). If no gate is prevented from being added to a partial path (gates that are not in gate g_i's fan-in and fan-out cones are eliminated for the path generation for g_i), the complete paths generated from the path generation are the global longest paths.

The advantage of finding the global longest paths is as follows: If there are p global longest paths covering gate g_i, these paths must be the p longest paths through g_i. Therefore, at the beginning of path generation, global longest path generation can cover many gates. For comparison, if no global longest path generation were performed, gate g_i would also get some potential longest path through it during the path generation process for other gates, but in most cases, they would not be "verified" until the path generation for g_i is performed.

However, as the global longest path generation finds more paths, the possibility that more gates are covered falls. The worst case is that almost all the global long paths only pass through a small subset of the gates. Therefore, at the beginning the global longest path generation is useful, but after a certain number of paths, it is necessary to apply the path generation targeting individual gates.

In this work, a two-phase strategy is used: Run the global longest path generation until no more gates benefit. Then, run the path generation for individual gates that are not fully covered during the global path generation.

The benefits from the global longest path generation are not only that it drops some gates from the individual path generation, but also that it speeds up the individual path generation for the gates that are not dropped. Suppose the length of the last generated global longest path is L. During the individual path generation, all partial paths with min esperance greater than L can be removed because all the testable paths whose length is greater than L have already been generated. This technique is especially useful when a circuit contains many long false paths.

2.3 Experimental Results for Combinational Circuits

We performed experiments on the ISCAS85 and enhanced scan versions of the ISCAS89 benchmark circuits. In our experiments, $K = 1$ means the path generation tries to find the longest path (one path) with either a rising or falling transition on the target gate output. It does not cover both slow-to-rise and slow-to-fall transition faults for the target gate output. Table 2.1 shows the results for generating the five longest paths through each gate ($K = 5$) under the robust and nonrobust sensitization criteria. Under each criterion, the number of testable paths generated and the execution time are listed. The

TABLE 2.1

Path Generation Results for Generating the Five Longest Paths ($K = 5$) through Each Gate

		Robust		Nonrobust	
Circuit	Number of Gates	Number of Testable Paths	CPU Time (m:s)	Number of Testable Paths	CPU Time (m:s)
c432	160	235	0:01	252	0:01
c499	202	423	0:01	430	0:01
c880	383	737	0:02	737	0:02
c1355	546	828	0:04	835	0:06
c1908	880	1,287	0:25	1,269	0:33
c2670	1,269	2,124	0:13	2,251	0:19
c3540	1,669	2,764	0:58	2,796	0:56
c5315	2,307	3,921	0:26	4,101	0:23
c6288	2,416	3,679	38:10	3,458	32:29
c7552	3,513	5,462	0:59	5,815	1:16
s9234	5,597	5,002	1:17	5,283	1:31
s13207	7,951	7,643	1:52	7,711	2:10
s15850	9,772	8,957	2:39	9,377	2:31
s38417	22,179	25,175	6:43	29,446	8:09
s38584	19,253	28435	10:49	31,095	11:23

size of the path store was set to 3,000, which is rarely reached. The number of gates in each circuit is given in the second column in the table. Clearly, the upper bound of the total number of generated paths is K times the number of gates in the circuit. It can be seen that the actual number is much lower than the upper bound, indicating that many gates share paths. For all the circuits except c6288, the five longest testable paths through each gate were found, or it was proven that the number of testable paths through a gate was less than K (e.g., there was no transition fault test for the gate), or the number of structural paths through the gate was less than $K/2$. For circuit c6288, our algorithm aborted on 20 gates (of 2,416 gates) that have transition tests.

Table 2.2 shows the execution time if some of the implicit false path elimination techniques are not used, assuming $K = 5$ and the robust sensitization criterion are used. The results assuming that forward trimming or S-PERT is not used are listed in columns 2 and 3. Column 4 lists the execution time if neither the relations between gates nor the global longest path generation is used. Column 5, using all the techniques, is copied from Table 2.1 for comparison. It can be seen that the forward trimming technique significantly reduces the execution time. The reason is that, in most cases, the local conflicts do not cause an exponential number of false paths, and these local

TABLE 2.2

Execution Time Comparison, Assuming Some of the
Implicit False Path Elimination Techniques Are Not Used
($K = 5$, Robust)

Circuit	CPU Time (m:s)			
	No FT	No SP	No R&G	Table 2.1
c432	0:01	0:01	0:02	0:01
c499	0:02	0:01	0:01	0:01
c880	0:04	0:02	0:03	0:02
c1355	0:08	0:04	0:08	0:04
c1908	2:25	0:23	0:36	0:25
c2670	0:21	1:07	0:43	0:13
c3540	3:49	0:57	1:45	0:58
c5315	0:37	0:26	0:46	0:26
c6288	58:13	X	62:38	38:10
c7552	1:21	0:55	1:42	0:59
s9234	1:48	1:11	5:04	1:17
s13207	2:24	2:49	4:13	1:52
s15850	3:12	2:33	6:25	2:39
s38417	8:20	6:35	15:26	6:43
s38584	14:31	10:42	20:09	10:49

FT, forward trimming; R&G, neither the relations between gates
nor the global longest path generation is used; SP, S-PERT;
X, the path generation did not finish within 12 h.

conflicts can be efficiently removed by forward trimming. S-PERT is not useful for most circuits, but it significantly helps c2670 and c6288, which have many reconvergences and long untestable paths. Without it, c6288 cannot finish path generation in a reasonable time (12 h).

If neither relations between gates nor global longest path generation is used, the path generation slows significantly for some circuits. But, if only one of the techniques is not used, only a 5–10% slowdown is observed. This phenomenon indicates that the benefits from the two techniques greatly overlap for most circuits. If we look at c2670 and c6288, it can be found that most long structural paths are untestable. If the global longest path generation is applied, these untestable paths are recognized, and during the path generation for a particular gate, none of these paths is considered. Similarly, if the relations between gates are applied, these long untestable paths are also recognized only once. For a short untestable path p, the effects are limited because this path may be a long path through gate g_i, assuming all the paths through g_i are short paths, but it is not likely to be a long path through another gate g_j. Thus, during the path generation for g_j, it is possible that the K longest testable paths through it have already been found before considering path p. Therefore, the more untestable global long paths are, the more benefits from applying the two techniques.

Another interesting phenomenon is that the forward trimming technique is more efficient for the ISCAS85 circuits, while relations between gates and global longest path generation are more efficient for the ISCAS89 circuits. The reason is that in most ISCAS89 circuits, more than half of the gates are inverters and buffers. Since these gates have only one input, no path is blocked if a logic value is assigned to their inputs. Thus, the efficiency of the forward-trimming technique decreases. However, the efficiency of the relations between gates and global longest path generation increases because the inverters and buffers have only one fan-in gate. If the K longest testable paths have been found for the only fan-in gate and all these paths go through the inverter or buffer, which is very likely, then the path generation for this inverter or buffer can be skipped.

From the analysis, it can be concluded that the forward-trimming and S-PERT techniques reduce the search space and guide the path generation more accurately to generate a testable path, while the relations between gates and global longest path generation mainly help avoid repeated work.

Table 2.3 shows the execution time as K is increased. It can be seen that the execution time is sublinear in K. For c7552, for example, there are 1,196 paths generated when $K = 1$ and 19,888 paths when $K = 20$. The number of generated paths when $K = 20$ is more than 16 times that of $K = 1$, but the execution time increases less than 87%. The reason is that when the first longest path through a gate is generated, in the path store there are many partial paths almost reaching a PO. Therefore, the cost of generating more paths is small. Sublinear CPU time growth has been observed up to $K = 500$.

TABLE 2.3

Execution Times for Different *K* Values (Robust)

	CPUTime(m:s)			
Circuit	*K*=1	*K*=5	*K*=10	*K*=20
c432	0:01	0:01	0:01	0:02
c499	0:01	0:01	0:01	0:01
c880	0:02	0:02	0:03	0:03
c1355	0:04	0:04	0:05	0:06
c1908	0:22	0:25	0:26	0:28
c2670	0:12	0:13	0:14	0:16
c3540	0:46	0:58	1:06	1:28
c5315	0:22	0:26	0:29	0:36
c6288	30:53	38:10	40:33	45:04
c7552	0:45	0:59	1:06	1:24
s9234	1:07	1:17	1:29	1:45
s13207	1:30	1:52	2:21	3:19
s15850	2:06	2:39	3:17	4:15
s38417	5:30	6:43	7:48	9:47
s38584	7:09	10:49	13:16	16:44

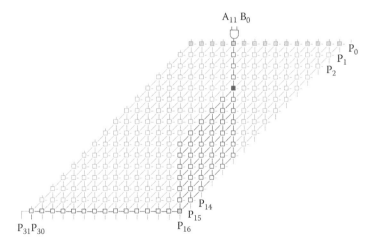

FIGURE 2.12
ISCAS85 circuit c6288 16 × 16 multiplier.

ISCAS85 circuit c6288 is a special case because it contains an exponential number of false paths. Figure 2.12 [Hansen 1999] shows the structure of circuit c6288, which is a 16 × 16 bit multiplier. The circuit contains 240 adders, among which 16 are half adders, which are shaded. $P_{31} \ldots P_0$ are the 32-bit outputs. Each floating line, including P_0, is fed by an AND gate, whose inputs are connected to two PIs. Figure 2.13 shows the structure of a full

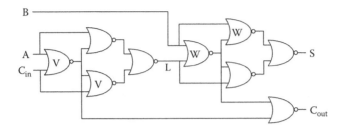

FIGURE 2.13
Full adder module in c6288 (schematic).

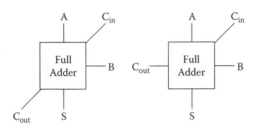

FIGURE 2.14
Full adder modules in c6288 (symbolic).

adder in c6288, and Figure 2.14 shows the block diagram. The 15 top-row half adders in Figure 2.12 lack the C_{in} input. Each of them has two inverters at locations V in Figure 2.13. The single half adder in the bottom row lacks the B input, and it has two inverters at locations W.

The longest structural paths through a particular gate or adder in c6288 (e.g., the black one in Figure 2.12) are highlighted. All these paths have equal length and include the longest structural paths within the adders. However, the longest structural paths from the input A or C_{in} to the output C_{out} in the adders are not robustly testable. This is the main feature that causes most false paths in this circuit. In our experiments, the S-PERT depth is set to 6, and this type of local conflict is identified in the preprocessing phase.

2.4 Extension to Scan-Based At-Speed Testing of Sequential Circuits

The KLPG algorithm described in the previous sections [Qiu 2003] is suitable for combinational or enhanced scan circuits that can apply two independent vectors to the logic. The flip-flops in a typical scan design have input multiplexers that are controlled by a *scan-enable* signal to switch between functional and scan operation. In this work, we assume that all flip-flops are clocked by a system clock, but the algorithm is readily extended to handle gated clocks. The typical scan design requires that the second test vector be derived from the first, so the combinational KLPG algorithm described in previous sections must be extended [Qiu 2004]. To apply input transitions to a typical scan design, two clocking techniques can be used.

The launch-on-shift [Savir 1992; Patil 1992] clocking technique is as follows:

1. The circuit is set to scan mode. The first test vector is scanned into the scan chains using the slow scan clock, and the values are set on PIs.

2. The second test vector is obtained by shifting the scan chain by one bit. Usually, the PIs do not change values to minimize the number of high-speed pins required on the tester.

3. The circuit is set to the functional mode by flipping the scan-enable signal and pulsing the system clock to capture the circuit values in the flip-flops. The values on POs are usually not captured to reduce the number of high-speed tester pins.

4. The circuit is set to scan mode, and the values in the scan chains are scanned out using the slow scan clock. This step can be overlapped with step 1.

The advantage of this approach is that test generation for combinational circuits can be applied without many modifications. Scanned flip-flops are considered PIs in test generation for combinational circuits, and the adjacent scan bit dependencies can be added to the search procedure. The disadvantage of the launch-on-shift technique is that the scan-enable signal must operate at full speed, so this technique is restricted to slower-speed circuits.

The *launch-off-capture* [Savir 1994a] clocking technique is as follows:

1. Same as step 1 of the launch-on-shift technique.

2. The circuit is set to functional mode. Delay cycles are inserted into the test program as necessary until the scan-enable signal settles.

3. The system clock is pulsed twice. At the first clock pulse, the second test vector is derived from the first vector. At the second clock pulse, the test is performed, and the output values are captured in the scanned flip-flops. The values on POs are usually not captured to reduce the number of high-speed tester pins.

4. Step 4 is the same as for the launch-on-shift technique.

The advantage of this approach is that it does not require the scan-enable signal to operate at full speed. The disadvantage is that it typically requires more test generation effort and more test patterns to achieve high fault coverage.

2.5 Path Generation for Scan Circuits

The KLPG combinational circuit test generation algorithm must be extended to include the input and output constraints from the launch-on-shift/capture techniques and include time frame expansion for the launch-off-capture technique.

2.5.1 Implications on Scanned Flip-Flops

Direct implications can be performed on scanned flip-flops as well as regular gates to detect most local conflicts and eliminate sequential false paths. Since local conflicts are the fundamental reason for false paths in most circuits, performing direct implications as much as possible can identify most false paths and significantly speed up the test generation process.

If the launch-on-shift technique is used, the logic values on neighboring scan flip-flops are dependent on each other. For example, in Figure 2.15, the logic value of cell A in the first vector is the same as that of cell B in the second vector. The relation between cell B and C is the same. Therefore, if there is a rising transition assigned to cell B, direct implications would try to assign logic 1 to cell A in the first vector and logic 0 to cell C in the second vector and propagate the new assignments throughout the circuit. If there are any conflicts, the partial path is a sequential false path under the launch-on-shift constraints.

If the launch-off-capture technique is used, the second vector is the output of the combinational circuit, derived from the first vector, excluding the PI and PO bits. In other words, $V_2 = C(V_1)$, where V_1 and V_2 are the two vectors, and C is the logic of the combinational circuit. For example, if it is assumed that a testable path has a rising transition launching from cell A and a rising transition captured on cell B, in Figure 2.15, then for the first vector, output a'

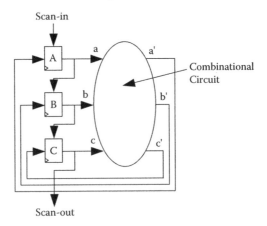

FIGURE 2.15
Implications on scanned flip-flops.

must be a logic 1 (then it becomes the value for input a in the second vector); for the second vector, input b must be a logic 0 because it is derived from the first vector. Then, more direct implications can be performed from a' and b.

2.5.2 Constraints from Nonscanned Memories

If the circuit is not a full scan, some nonscanned flip-flops may not be initialized after the first vector is scanned in. Flip-flops not controlled by the system clock are considered uncontrollable. Industrial designs also contain embedded memories, whose values cannot be easily initialized during a logic test. These values are considered "uncontrollable" during test generation.

Treating uncontrollable values the same as unknown values can result in low fault coverage. We use a 7-value algebra to distinguish unknown from uncontrollable values. The seven values are logic 0/1, x (unknown/unassigned), u (uncontrollable), 0/u (0 or uncontrollable), 1/u (1 or uncontrollable), and x/u (unknown or uncontrollable). At the beginning of test generation, the lines from the nonscanned memories are u, and all the other lines are x. Both u and x have "don't know" values, but x may be assigned a value in the test generation process (assuming it is controllable). Figure 2.16 shows the truth table of a 2-input AND gate in the 7-value algebra. For example, if one input is x and the other is u, the output is 0/u because if the input with x is assigned a logic 0, the output becomes 0, but if this input is assigned a logic 1, the output becomes uncontrollable. Figure 2.17 shows two examples, assuming M_1 is a nonscanned memory cell and M_2 is a scanned flip-flop. The logic values assigned by during test initialization are shown. If the conventional 3-value algebra (0, 1, x) is used, all the lines are assigned x's.

Using the 7-value algebra significantly speeds up the test generation because it divides unknown values into controllable and uncontrollable categories. In the example shown in Figure 2.17a, since the logic value on line n_3 can never be logic 1, all the paths through line n_4 are false. Thus, the test generation stops growing partial paths at line n_4, and all the gates in the fan-out cone of line n_4 are pruned. If the conventional 3-value algebra is used, the test generation may have to generate all the paths through line

	0	1	x	u	0/u	1/u	x/u
0	0	0	0	0	0	0	0
1	0	1	x	u	0/u	1/u	u
x	0	x	x	0/u	0/u	x/u	x/u
u	0	u	0/u	u	0/u	u	0/u
0/u	0	0/u	0/u	0/u	0/u	0/u	0/u
1/u	0	1/u	x/u	u	0/u	1/u	x/u
x/u	0	u	x/u	0/u	0/u	x/u	x/u

FIGURE 2.16
Truth table of an AND gate in 7-value algebra.

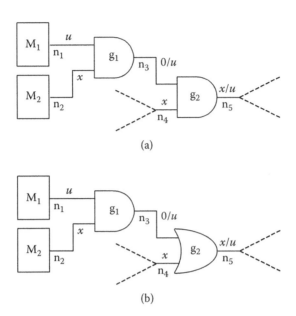

(a)

(b)

FIGURE 2.17
Application of 7-value algebra.

n_4 and find there is no test pattern for any of them. Moreover, by looking at the logic values on line n_5, it can be learned that it is impossible to intentionally make a transition on this line because logic 1 is not achievable; therefore, both slow-to-rise and slow-to-fall faults on this line are untestable. Since all the paths through line n_4 must contain line n_5, it can also be determined that both delay faults on line n_4 are untestable. In summary, many faults can be proven untestable by simple analysis before the test generation is performed.

Figure 2.17b shows another example. When a partial path reaches gate g_2, the value of the second vector on side input n_3 is set to logic 0 (noncontrolling value, for both robust and nonrobust tests). Then, direct implications are performed backward. The value of the second vector on line n_2 is set to 0, and further direct implications for the first vector can be performed from M_2. Thus, conflicts can be found earlier. If the conventional 3-value logic is used, direct implications stop at gate g_1, and some conflicts may remain hidden.

2.5.3 Final Justification

The final justification algorithm must be modified to support launch-on-shift and launch-off-capture techniques. Since the two vectors are dependent, whenever a decision (a logic value on any bit in either vector) is made at a PI or scanned flip-flop, direct implications have to be performed to trim the search space. For the launch-on-shift approach, both vectors can be justified in this way. For the launch-off-capture approach, because the second vector is derived as the circuit response to the first vector, one time frame expansion is used. The first vector V_1 can be generated within one time frame, but since the second vector $V_2 = C(V_1')$, the goal is to find a satisfying V_1'. Because V_1 and V_1' are identical excluding the don't care bits, in the justification process there must be no conflicts between V_1 and V_1'; that is, a bit is logic 1 in V_1 but 0 in V_1' (it is consistent if one of them is a don't care). Similarly, whenever a decision is made on any bit in either vector, direct implications must be performed to keep the logic assignments on any line in the two identical circuits consistent.

2.6 Experimental Results on Scan Circuits

Experiments were performed on the full-scan versions of the largest ISCAS89 benchmark circuits and two partial-scan industrial designs, *controller1* and *controller2*. The nominal timing models are used for computing path delays for *controller2*. The unit delay model is used for the ISCAS89 circuits and *controller1*.

2.6.1 Robust Test

Table 2.4 shows the results for generating the longest robustly testable path for each fault under the launch-off-capture and launch-on-shift constraints. It is assumed that at each fault site there are slow-to-rise and slow-to-fall delay faults. The number of faults is twice the number of lines in a circuit and the same as the number of transition faults. Column 3 shows the upper bound of faults that are detectable when holding inputs constant and masking outputs. Columns 4 and 5 show the number of PIs and scan flip-flops for each circuit. There are 38 nonscanned memory cells in *controller1* and 5,557 in *controller2*. Columns 6–8 show the results for the launch-off-capture approach, and columns 9–11 show those for the launch-on-shift approach. Columns 6 and 9 show the number of paths generated. Before test compaction, each generated path has a test pattern. The number of patterns after compaction is shown in columns 7 and 10. The test patterns are compacted using a simple greedy static compaction algorithm, in which each new pattern is combined with the first compatible existing pattern. Columns 8 and 11 show the CPU time.

2.6.2 Comparison to Transition Fault Tests

One goal of the KLPG test set is to have the same transition fault coverage as the transition fault test set. Because some lines in the circuit can only be

TABLE 2.4

Robust Test Generation Summary

Circuit	Number of Lines	UB Number of Detectable Faults	Number of Primary Inputs	Number of Scan Flip-Flops	Launch on Capture			Launch on Shift		
					Number of Paths Generated	Number of Test Patterns	CPU Time (m:s)	Number of Paths Generated	Number of Test Patterns	CPU Time (m:s)
s1423	1,423	2,420	20	74	395	215	0:13	666	191	0:07
s1488	1,488	1,310	11	6	192	87	0:01	206	81	0:01
s1494	1,494	1,324	11	6	193	85	0:02	204	79	0:01
s5378	5,378	7,564	38	179	1,799	406	0:07	1,110	94	0:04
s9234	9,234	16,166	39	211	2,376	790	3:59	3,608	681	2:52
s13207	13,207	22,886	65	638	3,220	909	2:25	6,469	1,635	1:03
s15850	15,850	24,338	80	534	2,637	472	2:35	5,828	645	1:08
s35932	35,932	59,246	38	1,728	9,762	36	14:31	12,194	44	8:15
s38417	38,417	74,926	31	1,636	14,905	949	14:21	17,554	655	2:46
s38584	38,584	59,454	41	1,426	9,723	526	11:20	21,047	679	4:28
controller1	86,612	130,692	38	3,503	12,275	2,275	130:10	19,626	657	102:41
controller2	1,966,204	1,815,222	201	57,352	493,779	70,670	132h[a]	714,116	43,289	57h

[a] A commercial ATPG tool took more than 48 h for transition fault test generation using the launch-off-capture technique for *controller2*.

tested by nonrobust or transition tests, we must construct a test set using all three test constraints. In this work, we target one longest-rising and one longest-falling path for each line, and term this test set KLPG-1.

The KLPG-1 test set is constructed as follows: First, a longest robust path is generated through a line. This may not be the longest path, but the quality of the test is guaranteed. If a robust test cannot be generated, the longest restricted nonrobust test is generated, if it exists. The following restrictions are placed on the existing path selection algorithm:

1. The path must be nonrobustly testable;
2. It must have the required transition at the fault site;
3. The local delay fault must be detected at a capture point if there is no other delay fault.

For example, path n_1-n_2-n_3-n_5 is a nonrobustly testable path in Figure 2.18. It is a valid nonrobust test for lines n_2 and n_3. However, it is not a valid nonrobust test for line n_5 because the glitch (or transition) may not happen if the delay of path n_1-n_4 is greater than the delay of path n_1-n_2-n_3 (violation of restriction 2). Similarly, it is not a valid nonrobust test for line n_1 because the slow-to-rise fault may not be detected even if there are no other delay faults (violation of restriction 3).

If a fault site does not have a nonrobust test, a transition fault test along a long path is generated using the path generation algorithm. This case usually happens when the local delay fault can only be activated or propagated through multiple paths, such as the slow-to-fall fault on line n_2 in Figure 2.19. The test quality is determined by the length of the shortest paths in the activating or propagating path set.

Table 2.5 shows a size comparison between KLPG-1 test sets and commercial transition fault test sets using the launch-off-capture approach. Columns 2–4 show the number of statically compacted robust/nonrobust/transition fault test patterns in the KLPG-1 test sets. Column 5 shows the total number

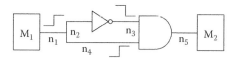

FIGURE 2.18
Restricted nonrobust test.

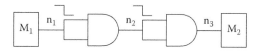

FIGURE 2.19
Transition fault test.

TABLE 2.5

Comparison of Test Size (Launch on Capture)

Circuit	Robust	Nonrobust	Transition	Total	Effective	Commercial
s1423	215	35	12	262	208	95
s1488	87	40	2	129	119	102
s1494	85	41	2	128	116	101
s5378	406	2	2	410	341	194
s9234	790	69	6	865	697	465
s13207	909	32	117	1,058	612	382
s15850	472	31	4	507	397	231
s35932	36	3	1	39	37	68
s38417	949	51	1	1,001	724	365
s38584	526	443	50	1,019	945	528
controller1	2,275	825	311	3,411	2,600	1,900
controller2	70,670	1,856	4,025	76,551	16,284	11,702

of compacted KLPG-1 patterns. If longest paths are not required, some KLPG-1 patterns can be dropped, but the transition fault coverage remains the same. Column 6 shows the number of transition fault effective patterns, which detect unique transition faults, out of the KLPG-1 test sets. The number of transition fault test patterns generated and dynamically compacted by a commercial tool is listed in column 7.

For most circuits, the KLPG-1 test sets are two to three times larger than the commercial transition fault test sets. For many circuits, the number of effective patterns is considerably smaller than the number of KLPG-1 patterns, indicating that many transition faults are easy to detect, but testing them through the longest paths requires more patterns. A dynamic compaction algorithm [Wang 2008b] and a low-cost fault coverage metric [Wang 2009] achieve KLPG-1 pattern counts comparable to a transition fault test.

2.6 Conclusions

We have described an algorithm that generates the K longest testable paths through each gate (KLPG) in combinational and scan-based sequential circuits. Test generation was achieved by growing partial paths from the PIs, and implicit false path elimination techniques were used to trim the search space and guide the search more accurately toward a testable path. Because many gates share long paths, the relations between gates and the global longest path generation heuristics efficiently reduce repeated work. A 7-value algebra was developed to handle nonscanned flip-flops and embedded memories whose logic values cannot be initialized during a scan test. Experimental results showed that the path generation is efficient, and it is the only tool that generates the longest testable path through each gate in circuit c6288.

References

[Bell 1996] J.A. Bell, Timing Analysis of Logic-Level Digital Circuits Using Uncertainty Intervals, MS thesis, Department of Computer Science, Texas A&M University, 1996.

[Benkoski 1990] J. Benkoski, E.V. Meersch, L.J.M. Claesen, and H.D. Man, Timing verification using statically sensitizable paths, *IEEE Transactions on Computer-Aided Design*, 9(10), pp. 1073–1084, October 1990.

[Chang 1993] H. Chang and J.A. Abraham, VIPER: An efficient vigorously sensitizable path extractor, ACM/IEEE Design Automation Conference, 1993, pp. 112–117.

[Fuchs 1991] K. Fuchs, F. Fink, and M.H. Schulz, DYNAMITE: An efficient automatic test pattern generation system for path delay faults, *IEEE Transactions on Computer-Aided Design*, 10(10), pp. 1323–1355, October 1991.

[Fuchs 1994] K. Fuchs, M. Pabst, and T. Rossel, RESIST: A recursive test pattern generation algorithm for path delay faults considering various test classes, *IEEE Transactions on Computer-Aided Design*, 13(12), pp. 1550–1562, December 1994.

[Goel 1981] P. Goel, An implicit enumeration algorithm to generate tests for combinational logic circuits, *IEEE Transactions on Computers*, C-30(3), pp. 215–222, March 1981.

[Gupta 2004] P. Gupta and M. Hsiao, ALAPTF: A new transition fault model and the ATPG algorithm, in *Proceedings IEEE International Test Conference*, 2004, pp. 1053–1060.

[Hansen 1999] M. Hansen, H. Yalcin, and J.P. Hayes, Unveiling the ISCAS-85 benchmarks: A case study in reverse engineering, *IEEE Design & Test of Computers*, 16(3), pp. 72–80, July–September 1999.

[Li 1989] W.N. Li, S.M. Reddy, and S.K. Sahni, On path selection in combinational logic circuits, *IEEE Transactions on Computer-Aided Design*, 8(1), pp. 56–63, January 1989.

[Lin 1987] C.J. Lin and S.M. Reddy, On delay fault testing in logic circuits, *IEEE Transactions on Computer-Aided Design*, 6(9), pp. 694–701, September 1987.

[Majhi 2000] A.K. Majhi, V.D. Agrawal, J. Jacob, and L.M. Patnaik, Line coverage of path delay faults, *IEEE Transactions on Very Large Scale Integration [VLSI] Systems*, 8(5), pp. 610–614, 2000.

[Murakami 2000] A. Murakami, S. Kajihara, T. Sasao, R. Pomeranz, and S.M. Reddy, Selection of potentially testable path delay faults for test generation, in *Proceedings IEEE International Test Conference*, 2000, pp. 376–384.

[Patil 1992] S. Patil and J. Savir, Skewed-load transition test: Part II, coverage, in *Proceedings IEEE International Test Conference*, 1992, pp. 714–722.

[Pomeranz 1995] I. Pomeranz, S.M. Reddy, and P. Uppaluri, NEST: A nonenumerative test generation method for path delay faults in combinational circuits, *IEEE Transactions on Computer-Aided Design*, 14(12), pp. 1505–1515, December 1995.

[Qiu 2003] W. Qiu and D.M..H. Walker, An efficient algorithm for finding the *K* longest testable paths through each gate in a combinational circuit, in *Proceedings IEEE International Test Conference*, 2003, pp. 592–601.

[Qiu 2004] W. Qiu, J. Wang, D. Walker, D. Reddy, X. Lu, Z. Li, W. Shi, and H. Balachandran, *K* longest paths per gate [KLPG] test generation for scan-based sequential circuits, in *Proceedings IEEE International Test Conference*, 2004, pp. 223–231.

[Savir 1992] J. Savir, Skewed-load transition test: Part I, calculus, in *Proceedings IEEE International Test Conference*, 1992, pp. 705–713.

[Savir 1994a] J. Savir and S. Patil, Broad-side delay test, *IEEE Transactions on Computer-Aided Design*, 13(8), pp. 1057–1064, August 1994.

[Shao 2002] Y. Shao, S.M. Reddy, I. Pomeranz, and S. Kajihara, On selecting testable paths in scan designs, in *Proceedings IEEE European Test Workshop, Corfu, Greece*, 2002, pp. 53–58.

[Sharma 2002] M. Sharma and J.H. Patel, Finding a small set of longest testable paths that cover every gate, in *Proceedings IEEE International Test Conference*, 2002, pp. 974–982.

[Smith 1985] G.L. Smith, Model for delay faults based upon paths, in *Proceedings IEEE International Test Conference*, 1985, pp. 342–349.

[Stewart 1991] R. Stewart and J. Benkoski, Static timing analysis using interval constraints, in *Proceedings IEEE International Conference on Computer-Aided Design*, 1991, pp. 308–311.

[Wang 2008b] Z. Wang and D.M..H. Walker, Dynamic compaction for high quality delay test, in *Proceedings IEEE VLSI Test Symposium*, 2008, pp. 243–248.

[Wang 2009] Z. Wang and D.M..H. Walker, Compact delay test generation with a realistic low cost fault coverage metric, in *Proceedings IEEE VLSI Test Symposium*, 2009, pp. 59–64.

3

Timing-Aware ATPG

Mark Kassab, Benoit Nadeau-Dostie, and Xijiang Lin

CONTENTS

3.1 Introduction

With fabrication processes moving well below 90 nm, manufactured devices are increasingly vulnerable to timing-related defects caused by resistive shorts or resistive opens, as well as to process variations, signal integrity issues, and so on. Delay testing has become a standard component of a scan-based logic test. The two most popular fault models used in the industry for a delay test are the transition fault model [Waicukauski 1987] and the path delay fault model [Smith 1985].

The transition fault model considers a gross delay at every gate terminal in the design and assumes that the additional delay at the fault site is large enough to cause a logic failure. Due to its limited fault population and ease of use, the transition fault model has been widely used in the industry. However, transition fault test generation ignores the actual delays through the fault activation and propagation paths and is more likely to detect a fault through a shorter path than the timing-critical one. As a result, the generated test set may not be capable of detecting small-delay defects that still result in positive slack along the shorter path targeted by pattern generation but not along longer paths.

The path delay fault model focuses on the testing of a set of predefined structural paths to detect the accumulated delays along these paths. Since the number of structural paths in design increases exponentially as the design size grows, it is infeasible to consider every path in the design explicitly. Instead, a subset of critical paths identified by static timing analysis tools is typically considered during test generation. Liou and coworkers [Liou 2003] proposed a statistical approach to select the critical paths for testing path delays. To achieve better delay fault coverage, another path selection approach [Sharma 2002] tried to cover every gate with the longest sensitized path through it. However, this approach suffered from high computational complexity for large designs, and it may not always be possible to test every gate based on robust or nonrobust conditions.

To benefit from the advantages of both the transition fault model and the path delay fault model, several mixed methods have been proposed. The segment delay fault model [Heragu 1996] targets the longest segments in the designs, and robust conditions are used to activate those segments. The as-late-as-possible transition fault (ALAPTF) model was proposed [Gupta 2004] to address the launch condition for the transition fault to catch small-delay defects accumulated along the transition launch segment. This approach does not consider the fault propagation path from fault site to observation point. Hence, the generated test pattern may still detect the faults through shorter paths, and the small-delay defects may not be detected. Moreover, finding the longest robustly tested launch segment is computationally complex.

The standard delay format (SDF) is an IEEE (Institute of Electrical and Electronics Engineers) standard to represent timing data used during the design process [IEEE 1497]. The data in the SDF file includes path delays, timing constraint values, and interconnect delays. Based on SDF data, all path delays in the design can be calculated. To detect small-delay defects by leveraging SDF, two approaches were developed: true-time delay (faster than at-speed) and timing-aware automatic test pattern generation (ATPG), also known as TAA.

True-time delay [Barnhart 2004; Uzzaman 2006] calculates the delays of the paths detecting each transition fault based on SDF data. Tests are then applied based on the lengths of the calculated paths, such that the test clock frequencies are higher than the system frequency. This enables the detection of small-delay variations by reducing the slack even along shorter paths. The downside is that this approach may lead to yield loss since small-delay variations may be within the available slack and are therefore redundant during system operation. In addition, when the test frequency is increased, paths with longer delays become multicycle paths and must be masked. The increase in the number of X states typically reduces the efficiency of on-chip scan compression methods and increases the scan data volume and test time.

Without changing test clock frequencies, TAA [Lin 2006] for the transition faults [Lin, 2001; Pomeranz, 2006] uses SDF data to guide the fault effect propagation and the fault site activation to detect the target fault through the longest path (i.e., the path with the least slack).

This chapter describes the test generation method used by TAA for transition faults. This chapter is organized as follows: In Section 3.2, the delay calculation with timing information and three delay test quality metrics used to measure the test set quality are described. Section 3.3 presents test generation and fault simulation for transition faults by integrating SDF data. Since TAA can generate significantly more test patterns than traditional ATPG, Section 3.4 presents approaches to trade off test quality and test cost. Section 3.5 reports the test generation results for industrial designs and discusses the test quality of the test sets generated by using different ATPG settings based on the delay test quality metrics described in Section 3.2.

3.2 Delay Calculation and Quality Metrics

Depending on the delay defect "size" (incremental delay modeled at a fault site), the extra delay introduced by the defect may not have an impact on all the paths passing though the fault site. The shorter paths passing through the fault site can tolerate more additional delay than the longer path passing through the same fault site due to larger slack margin. To achieve better test quality for small-delay defects, it is desirable to generate test patterns that propagate as many transition faults through corresponding longer paths as possible. Therefore, the delays of the paths activating and propagating the transition faults can be used to measure the delay test quality of different test sets.

3.2.1 Delay Calculation

To calculate the signal delays at each gate in a design accurately, timing-based simulation can be used to track the signal changes in time. (Process variations are not considered in this discussion; a single timing corner is used.) A simplified method was proposed [Carter 1987; Iyengar 1988a, 1988b], and it represents each signal by two logic values and two time numbers: the earliest possible arrival time of the initial value and the latest possible stabilization time of the final value. This method is still not easy or efficient to integrate into the test generation for target faults due to unspecified values that exist during ATPG. For timing-aware test pattern generation, the transition arrival and fault effect propagation times are used to approximate the path delay through a fault site in this chapter.

At a gate terminal z, the delay information associated with z is calculated using the following definitions:

- Transition arrival time to z, AT_z: The time to launch a transition at z from primary inputs or pseudoprimary inputs through the combinational logic.

- Fault effect propagation time from z, PT_z: The time to propagate the fault effect at z to a primary output or pseudoprimary output through combinational logic.

- Path delay through z, PD_z: The sum of AT_z and PT_z.

- Slack at z, S_z: The difference between the clock period and PD_z.

The formulas to calculate the transition arrival time from gate inputs to its output z are given next. They ignore the delay introduced by static hazards.

- When the transition at z changes from controlling value to noncontrolling value:

$$AT_z = \underset{i \in I_{cn}}{\text{MAX}}\left(AT_i + d_i\right) \tag{3.1}$$

where I_{cn} is the set of gate inputs with transitions from the gate's controlling value to its noncontrolling value, and d_i is the gate delay that propagates the transition from controlling value to noncontrolling value from the ith gate input to z.

- Otherwise:

$$AT_z = \underset{i \in I_t}{\text{MIN}}\left(AT_i + d_i\right) \tag{3.2}$$

where I_t is the set of gate inputs with transitions that imply the transition at z, and d_i is the gate delay that propagates the transition from the ith gate input to z.

Unlike the path delay model, which uses either robust or nonrobust conditions to evaluate the propagation time starting from a gate terminal through a specified path, the sensitization conditions used for propagation by the stuck-at and transition fault models are adopted here to evaluate the propagation time starting from a fault site. This method allows taking the reconvergent sensitization paths into account. As shown in Figure 3.1, the slow-to-fall transition fault at b can be detected by the pattern applied. However, neither a robust testable path nor a nonrobust testable path exists between b and z.

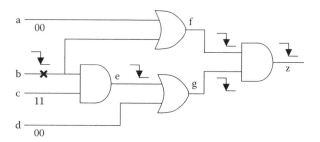

FIGURE 3.1
Transition propagating through reconvergent sensitization paths.

In fact, the transition at z is implied from b through reconvergent sensitization paths $b \to e \to g \to z$ and $b \to f \to z$. When calculating the propagation time from z to b, all the reconvergent sensitization paths between b and z are enumerated, and the propagation time at b is determined as follows:

$$PT_b = \underset{i \in S_{rs}}{\mathrm{MIN}}\left(PT_z + D_s\right) \tag{3.3}$$

where S_{rs} is the set of reconvergent sensitization paths including $b \to e \to g \to z$ and $b \to f \to z$, and D_s is the delay of the paths s in S_{rs}.

In general, the propagation time from the sensitized input i of a gate to its output z is calculated as follows:

$$PT_i = PT_z + d_i^v \tag{3.4}$$

where d_i is the gate delay to propagate the transition v at the input i through the gate, and v is a rising (falling) transition when the final value at i is logic 1 (0). When the gate output is a stem, the propagation time at the gate output is determined by the maximum propagation time among all of its branches.

Industrial designs may contain combinational loops. When calculating arrival time and propagation time for a design with combinational loops, the following criteria are used to avoid divergence during delay calculation:

- To calculate the arrival time at the gate z, every path from primary inputs or pseudoprimary inputs to z cannot contain the same gate more than once.

- To calculate the propagation time at the gate z, every path from z to primary outputs or pseudoprimary outputs cannot contain the same gate more than once.

- The path delay at the gate in a loop is not equal to the sum of the arrival time and the propagation time. It is equal to the delay of the longest path through the gate where the path cannot contain the same gate more than once.

3.2.2 Delay Test Quality Metrics

To evaluate the quality of a test set at detecting delay defects, the percentage of functional paths covered by the test set would ideally be computed. However, this method is impractical due to the exponential number of paths in the designs. To dramatically reduce computational complexity, the delay defect coverage is derived from evaluating the path delays through detected transition faults instead. Based on the derived path delays, three test quality metrics, delay test coverage (DTC), delay test quality coverage (DTQC), and statistical delay quality level (SDQL), are calculated to measure the delay test quality.

For each transition fault f, all three metrics require two types of path delay data through the fault site:

- Static path delay, PD_f^s: The longest path delay passing through f. It can be calculated through structural analysis of the combinational part of the design as an approximation for the longest functional path through f.
- Actual path delay, PD_f^a: The delay is associated with a test pattern t_i that detects f. It is defined as

$$PD_f^a(t_i) = AT_f + \underset{p \in P_s}{\text{MAX}}\left(PT_f^p\right) \tag{3.5}$$

where P_s is all of the sensitization paths starting from f. For a test set T, the actual path delay is defined as

$$PD_f^a = \underset{t_i \in T_D}{\text{MAX}}\left(PD_f^a(t_i)\right) \tag{3.6}$$

where T_D is the set of test patterns in T that detect f. When T_D is empty, PD_f^a is equal to 0.

3.2.2.1 Delay Test Coverage

The DTC [Lin 2006] is used to evaluate the effectiveness of a test set at detecting transition faults through the longest paths. DTC is independent of clock frequencies and the process-specific delay defect distribution. DTC is defined as follows:

$$W_f^{DTC} = \frac{PD_f^a}{PD_f^s} \tag{3.7}$$

$$DTC = \frac{\sum\limits_{f \in F} W_f^{DTC}}{N} \tag{3.8}$$

where F is the set of transition faults, and N is number of faults in F.

The upper bound of the DTC is equal to the transition fault test coverage. The closer the DTC reaches to the transition fault test coverage, the better the test set is at detecting transition faults through the longest paths.

3.2.2.2 Delay Test Quality Coverage

The DTC measures test quality by weighting each detected transition fault based on the path delay ratio of the actual path detecting the fault and the longest static path passing through the fault site. Since the test clock frequencies are not taken into account, it will give the same DTC for the same pattern set applied at different test clock frequencies. Since the smaller delay defects are more likely to occur than the larger delay defects and are more difficult to detect, it is desirable to measure the test set quality accounting for the delay fault size. To achieve this goal, the DTQC is defined as follows:

$$W_f^{DTQC} = \frac{T - \left(PD_f^s - PD_f^a\right)}{T} \tag{3.9}$$

$$DTQC = \frac{\sum_{f \in F} W_f^{DTQC}}{N} \tag{3.10}$$

where T is the clock period, F is the set of transition faults, and N is the number of faults in F.

Similar to DTC, the upper bound of the DTQC is equal to the transition fault test coverage. However, the DTQC tries to avoid biasing the quality of a test set by the faults along shorter static paths. If a fault is located on a path whose delay is closer to the clock period, the weight will become smaller when the fault is detected through shorter paths.

3.2.2.3 Statistical Delay Quality Level

To evaluate the quality of a test set at detecting delay defects, the statistical delay quality model (SDQM) [Sato 2005a] assumes that the delay defect distribution function $F(s)$ for a given process is known, where s is the defect size (incremental delay caused by the defect). Based on the simulation results of a test set, the detectable delay defect size for a fault f is calculated as follows:

$$T_f^{det} = \begin{cases} T_{TC} - PD_f^a & \text{if } PD_f^a > 0 \\ \infty & \text{if } PD_f^a = 0 \end{cases} \tag{3.11}$$

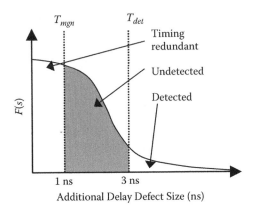

FIGURE 3.2
Delay defects escaped during testing.

where T_{TC} is the test clock period. The delay test quality metric, called the SDQL [Sato 2005a, 2005b], is calculated by multiplying the distribution probability for each defect as follows:

$$T_f^{mgn} = T_{SC} - PD_f^s \qquad (3.12)$$

$$SDQL = \sum_{f \in F} \int_{T_f^{mgn}}^{T_f^{det}} F(s)\,ds \qquad (3.13)$$

where T_{SC} is the system clock period, and F is the fault set. The motivation of the SDQL is to evaluate the test quality based on the delay defect test escapes shown in the shadow area in Figure 3.2 [Sato 2005b]. The lower the SDQL, the better the test quality achieved by the test set will be since the faults are detected with smaller actual slack.

3.3 Deterministic Test Generation

To generate a test set that detects transition faults through the longest paths, the existing transition fault ATPG process has to be enhanced using design timing data to guide test generation and measure the delay test quality metrics during fault simulation. This section describes both test generation and fault simulation with timing data in detail.

3.3.1 Test Generation with Timing Data

Generating a test that detects a target transition fault includes two steps:

- Step 1: Fault effect propagation: The activated transition fault effect is propagated to a primary output or a pseudoprimary output through the longest path. If the state element driven by the pseudoprimary output is not an observation point, the fault effect is further propagated through the state element until an observation point is reached. The delay of the propagation path during this second phase is not important.

- Step 2: Justification: All of the unjustified gates required for fault effect propagation and fault activation after the fault propagation path has been created in the first step are justified. When the unjustified gates are used to control the activation conditions at the fault site, they are justified by using choices that maximize the transition arrival time at the fault site.

The steps proposed do not guarantee that the fault is detected through the absolute longest path since it maximizes the fault propagation path and the fault activation path independently. When considering the extremely high computational complexity to maximize the sum of arrival time and fault propagation time, the proposed algorithm achieves a reasonable trade-off between test generation effort and finding the longest path passing through the fault site.

To guide the fault effect propagation and gate justification, a preprocessing step is carried out to calculate the longest static propagation times and the longest static arrival times for each gate in the design before test generation. When propagating the fault effect at a stem gate further, the branch with the longest static propagation delay is preferred to maximize the fault propagation time. When justifying the gates controlling the fault activation conditions, the choice with the longest static arrival time is selected first to maximize the transition arrival time to the fault site.

Although using the longest static propagation time at the stem creates the longest sensitization path, it may potentially propagate the fault effects through the same fault propagation paths repeatedly while targeting different faults. Besides testing the faults through the longest paths, covering a wide range of paths should be another requirement for the test generation. To achieve this goal and simultaneously compensate for imperfect SDF data, a weighted random method [Lin 2007b] can be used to select the branches for fault propagation. Assuming that the fault effect is going to be propagated from a stem with n branches, the probability of selecting the branch i for fault propagation is determined as follows:

$$w_i = \frac{PT_i^s}{\sum\limits_{i=1}^{n} PT_i^s} \qquad (3.14)$$

where PT_i^s is the longest static propagation time from the branch i. The weighted random method gives high probability to the paths with longer delays but does not ignore the paths with shorter delays.

Similar to the fault propagation, the weighted random method is also used to select the choices to justify the gates controlling the launch conditions at the fault site. The probability of selecting a justification choice is as follows:

$$w_i = \frac{AT_i^s}{\sum\limits_{i=1}^{n} AT_i^s} \qquad (3.15)$$

where AT_i^s is the longest static arrival time to the ith choice to justify the gate, and n is number of choices to justify the gate.

3.3.2 Fault Simulation with Timing Data

To derive the delay test quality of a test set, the fault simulator has to be enhanced to calculate the actual path delays for each detected fault. Two algorithms were proposed [Sato 2005b]: a transition-based algorithm and a path-based algorithm.

In the transition-based algorithm, a detected fault f cannot be dropped from the fault list unless the difference between the static path delay and the actual path delay of f is not greater than the difference between test clock period and system clock period. Since the evaluation of the static path delay is not always accurate and is typically greater than the delay of the longest functionally sensitizable path, this method may suffer from high computational complexity because many faults will not be dropped through fault simulation. However, it provides an exact way to identify the sensitization paths to propagate the fault effects.

The path-based method is a non-fault-oriented method. The actual path delays are derived from good machine simulation of each test pattern. Starting from each observation point, the nonrobust conditions are used to trace the circuit backward to identify the faults detected at the observation points. The path-based method has lower fault simulation time than the transition-based method. However, it cannot identify all of the detected transition faults since faults detected through reconvergent sensitization paths cannot be found with nonrobust conditions.

In the experimental results, the path-based method is used to calculate the actual path delays. Unlike the method proposed [Sato 2005b], the sensitization conditions are used such that the reconvergent sensitization paths are identified during the backward tracing to avoid missing detected transition faults. The fault simulation procedure for a test pattern t_i is described next.

Procedure *fault_simulation(t_i)*

1. Execute good machine simulation for t_i.

2. Execute fault simulation for all undetected faults and drop a fault from the fault list if it is detected by t_i.

3. Starting from all the observation points, trace the good machine backward to identify sensitization paths and calculate the maximal propagation time for each gate in the sensitization paths.

4. For all detected faults in the original fault list, do

   (a) Pick next detected fault f.

   (b) If f is not in the sensitization paths or the activation condition for f is not met, go to Step 4(a).

   (c) Calculate the arrival time to f.

   (d) Calculate the actual path delay passing through f.

   (e) Update the maximum actual path delay through f if the current actual path delay is larger.

End

3.4 Trade-off between Test Quality and Test Cost

After a test pattern or a set of test patterns is generated, the traditional transition fault test generator simulates undetected faults to find the faults detected by chance and to drop them from the target fault list. This strategy significantly reduces the test generation time and the test pattern count. However, using detection as the only criterion to drop a fault does not meet the requirement for detecting delay faults through the longest paths since the path detecting the fault may not be the longest sensitizable path through the fault site.

In general, the longest sensitizable paths to detect each delay defect are unknown in advance, and the structure-based static path analysis only provides an upper bound for the longest delays passing through each gate. As a result, the deterministic test generator is the only means to identify the longest sensitizable paths through each fault site such that a detected fault cannot be dropped from the fault list unless it is targeted by the deterministic test generator or the actual delay through the fault site is equal to the longest static delay. Obviously, this fault-dropping criterion not only increases the test generation effort dramatically but also may generate significantly more test patterns than the traditional transitional fault test generator.

To control the test cost, it is desirable to reduce the test pattern inflation while minimizing the reduction in the effectiveness of the test set at detecting small-delay defects. This section presents two approaches to achieve this trade-off: dropping based on slack margin (DSM) and timing-critical faults.

3.4.1 Dropping Based on Slack Margin

The DSM approach is applied during fault simulation. When using the DSM criterion, a fault f is dropped from the target fault list after fault simulating a test pattern t_i if the following conditions are met:

- The test pattern t_i detects f.
- The actual path delay created by t_i satisfies the following equation list:

$$\frac{PD_f^s - PD_f^a}{T_{TC} - PD_f^a} < \delta \tag{3.16}$$

where T_{TC} is the test clock period, and is a real number between 0 and 1. The equation means that a fault is dropped when the difference between the actual slack of the path detecting f and the minimum static slack associated with f is less than a predefined limit.

The DSM criterion with $\delta = 1$ is equivalent to the traditional fault-dropping criterion. When $\delta = 0$, almost all the faults in the fault list need to be targeted by the deterministic test generator. It is worth pointing out that as smaller values of δ are chosen for test generation, both the test pattern counts and the test generation efforts increase.

3.4.2 Timing-Critical Faults

When a delay defect is present in a device, its impact on the normal operation of the design depends on the defect's location and delay fault size:

- If the delay fault size is greater than the slack of the shortest functional path S_{min}, the circuit will malfunction independent of the delay defect's location along the functional path. Test patterns generated based on the traditional transition fault can ordinarily detect this type of delay defect during at-speed testing.

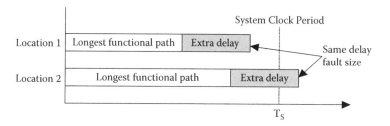

FIGURE 3.3
The impact of the delay defects on normal operation.

- If the delay fault size is less than the slack of the longest functional path S_{max}, the circuit will operate correctly even if the delay defects are present in the functional paths.

- When the delay fault size is between S_{min} and S_{max}, the circuit will malfunction during normal operation if the sum of the longest path delay passing through the defect and the extra delay introduced by the defect is greater than the system clock period; otherwise, the circuit will still operate correctly. This scenario is shown in Figure 3.3, where the delay defect does not have an impact on system operation if it is at location 1 but makes the circuit malfunction if it is at location 2.

Although it is important to test every functional path thoroughly to avoid delay defects escaping the test, this test strategy is impractical from a test cost point of view. Typically, small-delay defects have a higher probability of being introduced during manufacturing than large-delay defects [Nigh 2000]. The paths with tight slack margins are more vulnerable to violating timing constraints in the presence of the small-delay defects. As a result, a trade-off between test cost and test quality is to spend more test resources on testing the delay defects along timing-critical paths. Based on this observation, we can apply the following approach [Lin 2007a] to perform transition fault test pattern generation:

- Identify the transition faults located along the timing-critical paths, that is, the paths with tight slack margin, and target them explicitly using timing data to detect the faults through the longest paths.

- For faults not on timing-critical paths, use the traditional transition test generation strategy to detect them.

There are multiple ways to classify a fault as timing critical. In this chapter, we say a fault f is a timing-critical fault if it meets the following condition:

$$\frac{T_{TC} - PD_f^s}{T_{TC}} < \lambda \qquad (3.17)$$

where PD_f^s is the longest static path delay through f based on structural analysis, T_{TC} is the test clock period, and λ is a real number between 0 and 1. Equation 3.17 means that a fault is located along a timing-critical path when the minimum slack margin is less than some predefined limit. It is worth pointing out that a timing-critical transition fault f may be testable even if all timing-critical paths passing through f are untestable.

To generate a test set based on the timing-critical fault approach, one may generate two test sets independently: one for timing-critical transition faults while taking timing data into account and the other for non-timing-critical transition faults by using the traditional transition fault test generator. This method will increase the overall test pattern count since the first test set may detect some non-timing-critical faults. Therefore, we use the following test generation flow instead:

- During test generation, timing data are always used to guide the test generator that detects the target fault through the longest path.
- During fault simulation, a non-timing-critical transition fault is dropped from the fault list once it is detected. A timing-critical transition fault is dropped from the fault list only when the test generator has targeted it explicitly or the DSM criterion is met.

Since the majority of non-timing-critical transition faults are detected and dropped during fault simulation and only a small percentage of non-timing-critical transition faults are targeted explicitly by the test generator, the test generation flow proposed should have minimal impact on the test pattern count used for detecting the non-timing-critical faults even if the timing information is used during test generation.

3.5 Experimental Results

Two industrial designs were used to evaluate TAA. The characteristics of the designs are shown in Table 3.1. To calculate SDQL, the delay defect distribution must be available. Here, we use the same delay defect distribution function $F(s)$ as the one defined by Sato and coworkers [Sato 2005b]:

$$F(s) = 1.58 \times 10^{-3} \times e^{-2.1s} + 4.94 \times 10^{-6}. \tag{3.18}$$

For each design, three types of experiments were performed to generate three groups of broadside transition fault test sets:

TABLE 3.1

Characteristics of the Design

Design	# Gates	# Scan Cells	# Clocks	# Faults	SDQL
D1	94K	8.57K	2	225.8K	157,817
D2	2.43M	69.15K	8	4.74M	788,539

Gates, number of gates in the design; # Scan Cells, number of scan cells in the design; # Clocks, number of clocks in the design; # Faults, number of noncollapsed transition faults; SDQL, SDQL level before applying any test patterns.

- *Experiment 1*: Traditional ATPG with N-detection: Random decision order was used when launching the transition at a fault site and propagating the fault effect. A fault was dropped from the fault list once it was detected by the fault simulator N times.

- *Experiment 2*: TAA with DSM criterion: This applied the timing data from SDF to guide the test generation. The weighted random method described in Section 3.3.1 was used during test generation. A fault was dropped from the fault list when it was detected and the DSM criterion was met or when it was explicitly targeted by the test generator.

- *Experiment 3*: TAA with both DSM criterion and timing-critical fault identification: The setup was the same as in Experiment 2. However, the DSM criterion was only applied to faults identified as timing-critical faults. All non-timing-critical faults were dropped from the fault list once they were detected by the fault simulator. The distribution of the timing-critical faults based on Equation 3.17 in the two designs is shown in Figure 3.4. As shown in Figure 3.4, the percentage of timing-critical faults over the total number of transition faults was reduced as λ decreased. However, the distribution of the timing-critical faults was circuit dependent. In design D2, fewer than 6% of faults were timing critical when $\lambda = 0.15$. But, the percentage of timing-critical faults was more than 15% even if $\lambda = 0.05$ in design D1.

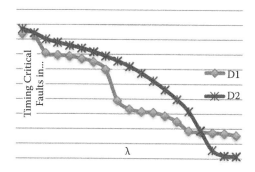

FIGURE 3.4
Distribution of timing critical faults.

TABLE 3.2

Test Generation Results

Design	ATPG Setting	Number of Patterns	TC (%)	DTC (%)	DTQC (%)	SDQL
D1	*Traditional ATPG*					
	$N = 1$	3,479	99.12	79.71	85.98	2,720.67
	$N = 3$	9,314	99.45	82.83	87.97	1,864.09
	$N = 5$	15,195	99.45	84.22	88.75	1,799.45
	$N = 10$	29,509	99.45	86.11	89.77	1,725.31
	Timing-Aware ATPG with DSM Criterion					
	$\delta = 1$	4,021	99.28	82.19	87.41	2,258.49
	$\delta = 0.5$	7,945	99.41	84.97	89.36	1,861.42
	$\delta = 0.25$	10,357	99.42	88.06	90.75	1,713.28
	$\delta = 0$	16,298	99.45	89.27	91.21	1,624.44
	Timing-Aware ATPG with Both DSM Criterion ($\delta = 0$) and Timing-Critical Faults					
	$\lambda = 0.15$	7,473	99.39	85.11	89.40	1,882.92
D2	*Traditional ATPG*					
	$N = 1$	2,923	91.76	64.82	74.14	2,264.4
	$N = 3$	7,208	91.84	66.83	75.46	2,154.51
	$N = 5$	11,352	91.86	67.71	76.01	2,107.1
	$N = 10$	21,593	91.86	68.95	76.75	2,041.59
	Timing-Aware ATPG with DSM Criterion					
	$\delta = 1$	3,006	91.84	66.46	75.01	2,182.34
	$\delta = 0.5$	3,666	91.87	70.59	77.61	1,964.52
	$\delta = 0.25$	4,082	91.87	70.77	77.70	1,955.71
	$\delta = 0$	4,858	91.87	71.16	77.89	1,937.85
	Timing-Aware ATPG with Both DSM Criterion ($\delta = 0$) and Timing-Critical Faults					
	$\lambda = 0.15$	3,331	91.86	69.90	77.15	2,002.32

The test generation results for the two designs are shown in Table 3.2. In the first experiment, the number of detections was run up to 10. When applying the DSM criterion in the second experiment, δ was set to 1, 0.5, 0.25, and 0. In the third experiment, δ was set to 0, and λ was set to 0.15. It resulted in 17.39% and 5.78% of faults being classified as timing-critical faults in the designs D1 and D2, respectively. For each ATPG run, five numbers are reported in the table to show the test generation results: the number of generated test patterns, test coverage, DTC, DTQC, and SDQL.

As shown in Table 3.2, the traditional ATPG improved all three delay test quality metrics as the number of detections N increased, along with a significant increase in pattern count. For example, compared to $N = 1$, $N = 10$ improved the DTC in D1 and D2 by 6.4% and 4.13%, respectively, but it generated 8.48 and 7.39 times more test patterns. Although the TAA with the DSM criterion $\delta = 1$ used the same fault-dropping criterion as $N = 1$ (i.e., a fault was dropped from the fault list once it was detected), the generated test set not only included a similar number of test patterns as $N = 1$ but also improved the DTC by 3.12% and 1.64% in D1 and D2, respectively. As δ decreased, all three delay test quality metrics improved at the expense of generating more test patterns. However, the number of generated test patterns with $\delta = 0$ was only 4.68 and 1.66 times larger than $N = 1$ in D1 and D2, respectively. When comparing the DTC, the DTC achieved by $\delta = 0$ was 3.16% and 2.21% higher than $N = 10$ in D1 and D2, respectively. These results demonstrated that TAA was more effective in detecting small-delay defects than N-detection ATPG.

The effectiveness of using the DSM criterion is obvious in both designs. From Table 3.2, it can be seen that all three delay test metrics achieved by setting $\delta = 0.25$ in D1 and $\delta = 0.5$ in D2, respectively, were much better than those achieved by $N = 10$. However, the test pattern count only increased by 2.98 in D1 and 1.24 times in D2, respectively, compared to traditional ATPG with $N = 1$.

Similar to using the DSM criterion, the effectiveness of using timing-critical faults to achieve a trade-off between delay test quality and test pattern count can be seen in both designs. When setting $\lambda = 0.15$ and $\delta = 0$, the number of timing-critical faults was 17.39% and 5.87% of the total number of transition faults in D1 and D2, respectively, and the generated test set achieved better delay test quality than $N = 5$ in D1 and $N = 10$ in D2, respectively, while the test pattern count increased by 114.8% and 14% compared to traditional ATPG with $N = 1$.

To better illustrate the effectiveness of the TAA at detecting delay faults through longer paths, Figures 3.5 and 3.6 show the fault distribution differences for different ATPG settings against the traditional ATPG with $N = 1$ in D1 and D2, respectively.

Figures 3.5a and 3.6a show the distribution of the differences between the maximal static path delay and the maximal actual path delay through each transition fault detected by the traditional ATPG with $N = 1$. Rather than showing the distribution of the static and actual path delay differences for the test sets generated by TAA directly, Figures 3.5b–3.5h and 3.6b 3.6h show the actual path delay distribution differences between various ATPG settings and the traditional ATPG with $N = 1$. The more detected faults moved to the region with smaller differences, the better delay test quality a test set achieved. Obviously, the TAA with $\delta = 0$ achieved the best delay test quality in both designs. When comparing Figure 3.5f and Figure 3.5h, it can be seen that they have similar distributions. However, the number of faults with actual path delay differences close to 0 is higher in Figure 3.5h than in Figure 3.5f. This is more evidence that the TAA with $\lambda = 0.15$ propagated more timing-critical faults through longer paths than the TAA with $\delta = 0.5$.

It is worth pointing out that test coverage achieved by the N-detection in D1 was slightly higher than that achieved by the TAA when δ was greater

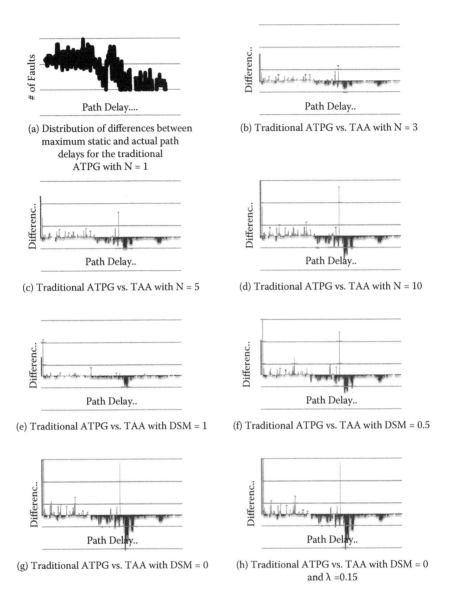

(a) Distribution of differences between maximum static and actual path delays for the traditional ATPG with N = 1

(b) Traditional ATPG vs. TAA with N = 3

(c) Traditional ATPG vs. TAA with N = 5

(d) Traditional ATPG vs. TAA with N = 10

(e) Traditional ATPG vs. TAA with DSM = 1

(f) Traditional ATPG vs. TAA with DSM = 0.5

(g) Traditional ATPG vs. TAA with DSM = 0

(h) Traditional ATPG vs. TAA with DSM = 0 and λ = 0.15

FIGURE 3.5
Path delay distributions of detected transition faults in design D1. (a) Distribution of differences between maximum static and actual path delays for the traditional ATPG with $N = 1$. (b) Traditional ATPG versus TAA with N = 3. (c) Traditional ATPG versus TAA with N = 5. (d) Traditional ATPG versus TAA with N = 10. (e) Traditional ATPG versus TAA with $DSM = 1$. (f) Traditional ATPG versus TAA with $DSM = 0.5$. (g) Traditional ATPG versus TAA with $DSM = 0$. (h) Traditional ATPG versus TAA with $DSM = 0$ and $\lambda = 0.15$.

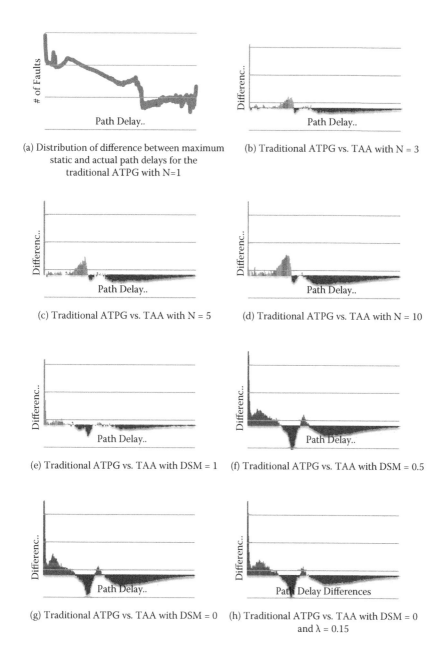

(a) Distribution of difference between maximum static and actual path delays for the traditional ATPG with N=1

(b) Traditional ATPG vs. TAA with N = 3

(c) Traditional ATPG vs. TAA with N = 5

(d) Traditional ATPG vs. TAA with N = 10

(e) Traditional ATPG vs. TAA with DSM = 1

(f) Traditional ATPG vs. TAA with DSM = 0.5

(g) Traditional ATPG vs. TAA with DSM = 0

(h) Traditional ATPG vs. TAA with DSM = 0 and λ = 0.15

FIGURE 3.6
Path delay distributions of detected transition faults in design D2. (a) Distribution of difference between maximum static and actual path delays for the traditional ATPG with $N = 1$. (b) Traditional ATPG versus TAA with $N = 3$. (c) Traditional ATPG versus TAA with $N = 5$. (d) Traditional ATPG versus TAA with $N = 10$. (e) Traditional ATPG versus TAA with $DSM = 1$. (f) Traditional ATPG versus TAA with $DSM = 0.5$. (g) Traditional ATPG versus TAA with $DSM = 0$. (h) Traditional ATPG versus TAA with $DSM = 0$ and $\lambda = 0.15$.

than 0.25. The main reason is that the design included some faults that were aborted by ATPG but accidently detected when filling unspecified bits randomly. Improving the ATPG algorithm, such as using SAT-based ATPG, may help to merge the test coverage gap.

References

[Barnhart, 2004] C. Barnhart, What is true-time delay test? *Cadence Nanometer Test Q* 1(2), 2004.

[Carter, 1987] J.L. Carter, V.S. Iyengar, and B.K. Rosen, Efficient test coverage determination for delay faults, in *Proceedings IEEE International Test Conference*, 1987, pp. 418–427.

[Gupta 2004] P. Gupta and M. Hsiao, ALAPTF: A new transition fault model and the ATPG algorithm, in *Proceedings IEEE International Test Conference*, 2004, pp. 1053–1060.

[Heragu, 1996] K. Heragu, J.H. Patel, and V.D. Agrawal, Segment delay faults: A new fault model, in *Proceedings IEEE VLSI Test Symposium*, 1996, pp. 32–39.

[IEEE 1497] Institute of Electrical and Electronics Engineers, *IEEE Standard for Standard Delay Format [SDF] for the Electronic Design Process*, http://www.ieeexplore.ieee.org/iel5/7671/20967/00972829.pdf

[Iyengar, 1988a] V.S. Iyengar, B.K. Rosen, and I. Spillinger, Delay test generation I—Concepts and coverage metrics, in *Proceedings IEEE International Test Conference*, 1988, pp. 857–866.

[Iyengar, 1988b] V.S. Iyengar, B.K. Rosen, and I. Spillinger, Delay test generation II—Concepts and coverage metrics, in *Proceedings IEEE International Test Conference*, 1988, pp. 867–876.

[Lin 2001] X. Lin, J. Rajski, I. Pomeranz, and S.M. Reddy, On static test compaction and test pattern ordering for scan design, in *Proceedings IEEE International Test Conference*, 2001, pp. 1088–1097.

[Lin 2006] X. Lin, K.-H. Tsai, C. Wang, M. Kassab, J. Rajski, T. Kobayashi, R. Klingenberg, Y. Sato, S. Hamada, and T. Aikyo, Timing-aware ATPG for high quality at-speed testing of small delay defects, in *Proceedings IEEE Asian Test Symposium*, 2006, pp. 139–146.

[Lin 2007a] X. Lin, M. Kassab, and J. Rajski, Test generation for timing-critical transition faults, in *Proceedings IEEEE Asian Test Symposium*, 2007, pp. 493–500.

[Lin 2007b] X. Lin, K.-H. Tsai, M. Kassab, C. Wang, and J. Rajski, Timing-aware test generation and fault simulation, U.S. Patent Application 20070288822, December 2007.

[Liou 2003] J.-J. Liou, L.-C. Wang, A. Krstic, and K.-T. Cheng, Experience in critical path selection for deep submicron delay test and timing validation, in *Proceedings of ASP-DAC*, 2003, pp. 751–756.

[Nigh 2000] P. Nigh and A. Gattiker, Test method evaluation experiments and data, in *Proceedings IEEE International Test Conference*, 2000, pp. 454–463.

[Pomeranz 1999] I. Pomeranz and S.M. Reddy, On N-detection test sets and variable N-detection test sets for transition faults, in *Proceedings IEEE VLSI Test Symposium*, 1999, pp 173–180.

[Sato 2005a] Y. Sato, S. Hamada, T. Maeda, A. Takatori, and S. Kajihara, Evaluation of the statistical delay quality model, in *Proceedings of Asia and South Pacific Design Automation Conference*, 2005, pp.305–310.

[Sato 2005b] Y. Sato, S. Hamada, T. Maeda, A. Takatori, and S. Kajihara, Invisible delay quality—SDQM model lights up what could not be seen, in *Proceedings International Test Conference*, 2005, pp. 1202–1210.

[Sharma 2002] M. Sharma and J.H. Patel, Finding a small set of longest testable paths that cover every gate, in *Proceedings IEEE International Test Conference*, 2002, pp. 974–982.

[Smith 1985] G.L. Smith, Model for delay faults based upon paths, in *Proceedings IEEE International Test Conference*, 1985, pp. 342–349.

[Uzzaman 2006] A. Uzzaman, M. Tegethoff, B. Li, K. McCauley, S. Hamada, and Y. Sato, Not all delay tests are the same—SDQL model shows true-time, in *Proceedings IEEE Asian Test Symposium*, 2006, pp. 147–152.

[Waicukauski 1987] J.A. Waicukauski, E. Lindbloom, B.K. Rosen, and V.S. Iyengar, Transition fault simulation, *IEEE Design and Test of Computer*, pp. 32–38, April 1987.

Section II

Faster-than-at-Speed

4

Faster-than-at-Speed Test for Screening Small-Delay Defects

Nisar Ahmed and Mohammad Tehranipoor

CONTENTS

4.1 Introduction

Technology scaling is introducing a larger population of timing-related defects in integrated circuits (ICs), and postsilicon performance verification is becoming an ever more challenging problem. The transition delay fault (TDF) model is widely practiced in industry to test such defects and is considered a cost-effective alternative to functional pattern generation [Saxena 2002; Lin 2003]. Functional pattern generation is practically infeasible for modern designs due to its intensive computational complexity, the large pattern count, and the very low fault coverage. Traditionally, TDF tests were generated assuming gross delay defect size and fixed cycle time equivalent to the functional operating frequency for each clock domain. Under such assumptions, a delay defect will be detected only when it causes a transition to reach an observation point (primary output or scan flip-flop) by more than the positive slack of the affected path. Slack of a path is a measure of how closely a transition on the respective path meets the timing to an observable point relative to the test cycle time.

Smaller technologies are introducing smaller-delay defect sizes, which may not necessarily cause a failure when detected through paths of any length. A delay defect that is not large enough to cause a timing failure under the fixed cycle time notion, regardless of the affected path length, is referred to as a *small-delay defect*. Small-delay defects may be caused by resistive opens and shorts. A small-delay defect might escape during a test if it is tested using a short path; the same defect might be activated on a longer path during functional operation, and it may cause a timing failure. Therefore, test coverage of small-delay defects on long paths determines the *quality* of test patterns as these might cause immediate field failures. Small-delay defects on short paths might become a *reliability* issue as the defect might magnify during subsequent aging in the field and cause a failure of the device. Hence, it is important to detect such defects during a manufacturing test using efficient techniques [Kruseman 2004].

Various techniques have been proposed in the past for improving the small-delay defect screening quality of a pattern set. A number of these methods, such as very low voltage (VLV) [Hao 1993] and burn in [Foster 1976], modify the operating conditions of the test environment and magnify the defect size, with the defects escaping at nominal conditions. However, in very deep submicron designs, the effectiveness of such methods is reduced due to reduction of supply voltage, increased cost. A new transition fault model, the as-late-as-possible transition fault (ALAPTF), was proposed [Gupta 2004]. The method tries to activate and propagate a transition fault at the target gate terminal through the least-slack path possible. Authors [Qiu 2004] have proposed a new automatic test pattern generation (ATPG) tool to generate K longest paths per gate for the transition fault test. The technique targets all the transition faults to find the longest paths. Another proposed technique [Kruseman 2004] is based on detecting a smaller delay on a shorter path by increasing the frequency of operation. The method groups a conventional delay fault pattern set into multiple pattern sets, which exercise almost equal-length paths. The different pattern sets are then applied at different frequencies to detect smaller delays. Because of increasing frequency, the capture edge might occur in the hazard region for some of the observation points and requires additional steps to mask the respective endpoints to avoid false timing failures.

To enhance the effectiveness of screening frequency-dependent defects, Lee and coworkers [Lee 2005] proposed a pattern selection methodology to reduce the delay variation of the selected pattern set, and a higher frequency is used for pattern application. The method uses a multiple-detect transition fault pattern set and statistical timing analysis techniques to reduce pattern delay variations. These methods are limited by the highest possible frequency of operation, which exacerbates the already-well-known issues of peak power during test and IR drop.

There is a growing industry concern for timing-aware ATPG tools. However, adding such intelligence to the tool comes at the expense of higher pattern volume and longer CPU (central processing unit) time as the tool has to perform extra processing to target each fault location through its longest path. Pattern generation procedures [Synopsys 2007b; Ahmed 2006b; Lin 2006] target the longest path through each fault site and utilize node slack information in the static timing analysis step during pattern generation. Although these techniques improve the detection of small-delay defects on long paths, short and intermediate paths still have enough slack to escape and require faster-than-at-speed test application. The Encounter True-Time Delay Test Tool™ [Cadence 2006] uses actual design timing (standard delay format, SDF) information to generate patterns for the faster-than-at-speed test. Although it still uses efficient ATPG algorithms and pseudorandom data to achieve high coverage in fewer patterns, it uses back-annotated timing information (SDF) to apply them faster than at speed and mask the longer paths at each higher frequency value to block transitions, reducing unpredictable responses. The pattern compression techniques are significantly affected by masking scan cells, whose path delay exceeds the faster-than-at-speed clock period.

Faster-than-at-speed test techniques [Cadence 2006; Kruseman 2004] for screening small-delay defects apply the patterns at higher than functional frequencies. While these techniques increase the test frequency to reduce the positive slack of the path, they exacerbate the already-well-known issue of IR drop and power during test. This may result in false identification of good chips as faulty due to IR drop rather than small-delay defects. A trivial solution to determine the maximum test frequency that a pattern can be applied would be to iteratively increase the applied frequency of the pattern on the tester until a good chip starts to fail. However, consideration of the test time impact and analysis required for a large test pattern set makes such a solution impractical. Also, it is impossible to apply each test pattern at an individual frequency either due to hardware limitations of the automatic test equipment (ATE) to generate higher frequencies or due to long synchronization times of on-chip clock generators (phase-locked loops, PLLs) affecting test time.

To address these issues, this chapter

- presents the practical issues of faster-than-at-speed delay test application.
- presents a case study of a design to illustrate the increase in both peak and average IR drop effects due to faster-than-at-speed pattern application. It is shown that the increase in IR drop directly relates to performance degradation due to effective voltage reduction reaching the gates in the circuit. Therefore, it is important to consider the performance degradation due to IR drop effects along with the positive slack when frequency is increased for small-delay fault detection.

- presents a novel faster-than-at-speed test technique for the application of transition fault patterns. The technique first divides the transition fault test patterns into different groups based on the maximum path delay affected in each pattern. It then performs worst-case IR drop analysis and estimates the related performance degradation in each group based on the switching cycle average power (SCAP) model [Ahmed 2007] and determines the maximum frequency of a pattern group.

The test method presented in this chapter reduces the risk of any false identification of good chips as faulty due to IR drop effects rather than small-delay defects. In addition, the method is timing aware and relatively fast for a large design.

4.2 Design Implementation

In this section, the physical synthesis of the case study design is briefly described. The design is a 32-bit processor core [Gaisler 2006; Ahmed 2010] with the following characteristics:

- 20K (24,000) bytes of memory
- Approximately 50K gates
- 4K scan flip-flops
- 124 IO (input/output) pads (53 bidirectional)

Scan-based test insertion was performed with eight scan chains [Synopsys 2007b]. During test mode, all the bidirectional pins are controlled in input mode to avoid any congestion problem. The memory modules are wrapped to provide controllability and observability of the outputs and inputs, respectively, during the scan test. The at-speed test methodology was to implement a functional launch-off-capture (LOC) transition fault test with a slow-speed scan-enable signal. The physical synthesis was performed by a place and route tool [Cadence 2004] using a 180-nm standard cell and IO (3.3-V) library [Cadence 2005]. The design was timing closed for an operating frequency of 100 MHz at a nominal operating voltage (1.8 V) and temperature (25°C) conditions. The scan shift path was closed for a lower frequency of 10 MHz. The power planning for the design was performed assuming a net toggle probability of 20% during functional operation.

Figure 4.1 shows the power/ground distribution network of the chip. Power and ground rings were created using higher-level routing layers (Metal5

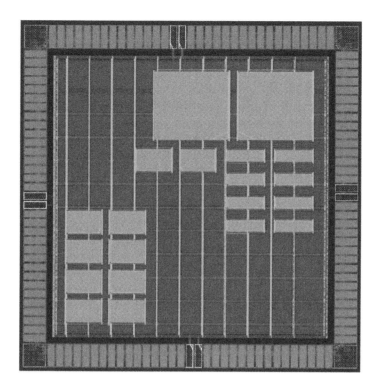

FIGURE 4.1
Design floor plan with power/ground distribution network [Ahmed 2010].

and Metal6), which supply power around the standard cell core area. Also, power and ground rings were created around each memory module. Four power (VDD) and ground (VSS) pads each were inserted, one on each side of the chip, and connected to the respective rings with wires referred to as trunks. After creating the power rings, the power and ground network was routed to the standard cells using horizontal and vertical straps. After the power distribution stage, the design was timing-driven placed and routed along with clock-tree synthesis and scan cell ordering to minimize scan chain wire length.

To determine an estimate of IR drop during functional operation, the design net parasitics (resistance and capacitance) were extracted using an extraction tool [Synopsys 2007b]. The average statistical IR drop using a vectorless approach was measured for both VDD and VSS nets considering 20% net toggle probability. The results showed a 3.7% voltage drop in the VDD network and a voltage bounce of 3.5% in the VSS network, which can be considered negligible. However, it is illustrated in the following sections that the actual IR drop during transition fault test pattern application was much higher compared to the measured statistical IR drop due to high switching activity resulting in performance degradation of the circuit.

4.3 Test Pattern Delay Analysis

IR drop is directly related to the switching activity and the time frame in which it occurs. Therefore, it is important to understand the path delay distribution of a pattern to identify potential high IR drop patterns. In this section, a detailed analysis of path delay distribution for a transition fault test pattern is presented, including dynamic IR drop effects, applied in two different cases: case 1 at a rated functional speed (i.e., at speed) and case 2 at a higher-than-functional frequency (i.e., faster than at speed) to detect small-delay defects.

The ATPG algorithms are based on zero-delay gate models, and the TDF test patterns are generated based on the ease of finding an affected path instead of a least-slack path through the target gate. Figure 4.2 shows the path delay distribution of a transition fault test pattern (*P1*) across all endpoints, generated using an ATPG [Synopsys 2007b]. The pattern was simulated using the design timing information in SDF. The SDF file was generated by a timing analyzer [Synopsys 2007b] using the parasitics extracted during physical synthesis at nominal operating conditions (25°C and 1.8 V).

An observation point at the end of a path (primary output or scan flip-flop) is referred to as an *endpoint*. Since the primary outputs are not observed due to insufficient timing accuracy of a low-cost tester to strobe at functional speed [Saxena 2002; Ahmed 2005], here an endpoint refers only to a scan flip-flop. An endpoint that does not observe a transition, referred to as a *nonactive* endpoint, is excluded in the path delay distribution discussed

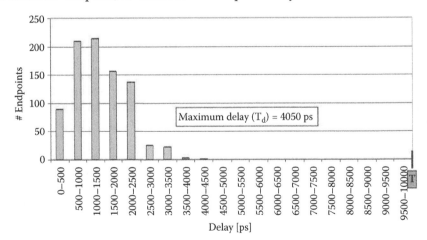

FIGURE 4.2
Path delay distribution across all endpoints for a single launch-off-capture transition fault test pattern (P1). As seen, most paths affected by this pattern are short, and a small-delay defect is likely to escape.

previously. It can be noticed that only a subset of endpoints observes transitions and, for this particular pattern, approximately 20% endpoints. These are referred to as *active* endpoints.

The functional clock period (f = 100 MHz) is represented by T = 10,000 ps. In this particular pattern set, the maximum path delay to an active endpoint was Td = 4,050 ps. Notice that a majority of the active endpoints fall in the range of 500–3,000 ps. This indicates that only a delay defect size *greater than half the clock cycle* can be detected on paths terminating at such endpoints. Note that the pattern delay analysis was performed with timing information at nominal operating conditions without taking IR drop effects into account.

4.3.1 Dynamic IR Drop Analysis at Functional Speed

The IR drop effects in the transition fault test pattern P1 applied at the rated functional frequency (f = 100 MHz) is referred to as the *inherent IR drop*. To measure the IR drop of the pattern, the switching activity inside the circuit was captured in the standard value change dump (VCD) format during gate-level timing simulation. The switching activity information (VCD file) along with physical design and technology library information is used by the system-on-chip (SOC) Encounter tool [Cadence 2004] to estimate the dynamic IR drop of the pattern. Figure 4.3 shows the average inherent IR drop plot on the power (VDD) network for pattern P1. The small dark region in black and white represents the area observing a voltage drop greater than 10% of VDD ($\Delta V \geq 0.18$ V).

To measure the performance degradation due to dynamic IR drop analysis, the average VDD and VSS voltage information of each instance in the design within the launch-to-capture window is stored. This information is then used to measure the cell delay degradation for each instance using the formulation

$$Scaled\_Cell\_Delay = Cell\_Delay * (1 + k\_volt * \Delta V)$$

where k_volt is a factor specified in the vendor-supplied technology library that accounts for the nonlinear delay scaling model, and ΔV is the effective voltage decrease. Here, a value of 0.9 ($1/V$) was used for k_volt, which means for a 6% effective cell voltage decrease (ΔV = 0.1 V), the cell delay increases by about 9% (i.e., $k\_volt * \Delta V$ = 0.09). Figure 4.4 compares the path delay distribution of the same pattern (P1) across all endpoints considering inherent IR drop effects and no IR drop effects. It can be seen that the distribution curve has been shifted to the right-hand side, which indicates slowing (performance degradation) of the cells due to IR drop effects. The maximum path delay to an active endpoint considering inherent IR drop effects increased to Td = 5,030 ps (21% delay increase over maximum path delay with no IR drop, 4,050 ps). However, there is still enough slack for small-delay defects to escape.

Note that, in this work, the timing library used was a nonlinear delay model (NLDM) library, and the ($k\_volt\_cell\_rise$) and ($k\_volt\_cell\_fall$) parameters

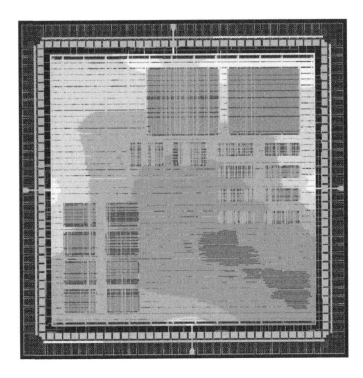

FIGURE 4.3
IR drop plot in VDD network for pattern P1 at rated functional frequency *f*. The dark region in black and white represents the area observing a voltage drop greater than 10% of VDD [Ahmed 2010].

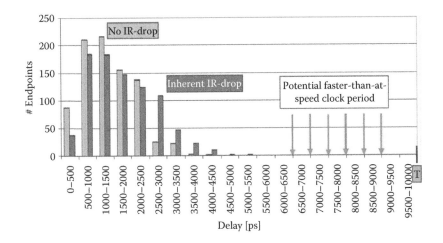

FIGURE 4.4
Comparison of path delay distribution across all endpoints for pattern P1 with no IR drop effects and inherent IR drop effects. The delay of paths in the presence of an IR drop has increased.

specified in the library, which determine the delay scaling, were used. Currently, the library used has only one set of such parameters. However, the delay scaling formulation provided in this chapter can be extended if the k factors are different for each library cell. Although it increases the accuracy of the analysis, it also would significantly increase the computational complexity.

To increase the small-delay defect screening capability, based on the maximum path delay Td, the test clock timing can be adjusted in the range $Td + \tau_{su} \leq T' < T$, where τ_{su} is the setup time of the scan flip-flop. Note that, for clarity, the timing margin for process variation effects is not shown in the equation, which can be easily incorporated. The potential faster-than-at-speed clock period ($T' = 1/f'$) for this pattern is as shown in Figure 4.4. Increasing the frequency improves the small-delay defect screening capability of the test pattern as the path delays affected by the pattern become relatively close to the clock period. However, applying a faster-than-at-speed test frequency increases the IR drop, which will have an impact on the performance of the design. In this particular case, the faster-than-at-speed clock period was selected to be $T' = 7,000$ ps ($f' \approx 145$ MHz), where $\tau_{su} = 200$ ps, and enough timing margin was added for process variation and dynamic effects such as IR drop and cross talk.

4.3.2 Dynamic IR Drop Analysis for the Faster-than-at-Speed Test

The IR drop in the VDD network for pattern P1 applied at the new selected higher frequency f' to detect small-delay defects is shown in Figure 4.5; it is referred to as the *faster-than-at-speed IR drop*. Note that the chip region observing a voltage drop greater than 10% VDD has increased significantly. As the test frequency is increased, the IR drop increases due to negative clock edge switching activity in the clock network toward the early cycle period compared to the rated functional frequency. This results in an increase of both peak and average IR drop due to faster-than-at-speed pattern application as shown in Figure 4.6.

To measure the voltage curve, the launch-to-capture window was split into 1-ns time frames, and the IR drop analysis tool was used to report the worst average IR drop in each time frame. Note that the IR drop effect is maximum in the beginning of the clock cycle due to high simultaneous switching activity, and it gradually decreases. The peak IR drop for VDD increased from 0.28 to 0.31 V (approximately by 10%), and the average IR drop during the switching activity time frame window considering both VDD and VSS ($V_{VDD,IR\text{-}drop} + |V_{VSS,IR\text{-}drop}|$) increased from 0.26 to 0.35 V (approximately 15%).

To measure the performance degradation, the pattern P1 was simulated again with a cell delay scaling technique (as explained in Section 4.3.1) based on each instance voltage obtained during the faster-than-at-speed IR drop analysis. The path delay analysis showed some interesting results; it was found that some of the endpoints that previously observed transitions or glitches were static. This might be because some of the reconverging

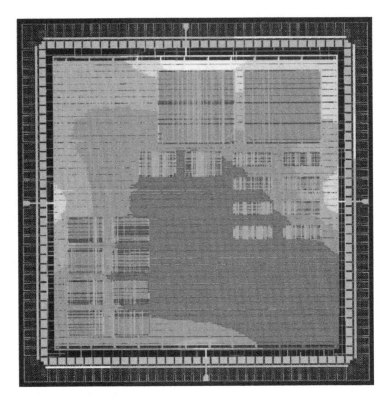

FIGURE 4.5
IR drop plot in VDD network for pattern P1 at faster-than-at-speed frequency ($f' = 1.4f$). Because of the use of a higher-than-functional frequency, a larger area observed a greater IR drop [Ahmed 2010].

paths were slowed due to IR drop effects, and the transitions/glitches were absorbed. Also, the TDF test pattern generation is a nonrobust technique, and fault detection is dependent on the off-path delays, unlike robust path delay fault detection. However, the maximum path delay ($Td = 5{,}250$ ps) was observed at the same endpoint, and it increased by only 4% compared to inherent IR drop. Further analysis showed that the longest path exercised by pattern P1 was only partially affected by the faster-than-at-speed IR drop, which explains the slight increase in the maximum path delay. In other words, the longest path affected by pattern P1 was only partially impacted by the IR drop; however, this may not likely occur in other patterns and may cause pattern failures due to IR drop effects rather than small-delay defects. Also, note that the faster-than-at-speed test frequency was selected based on the inherent IR drop pattern delay analysis with enough timing margin (almost 2 ns); this avoided the selection of even higher frequency leading to more adverse effects.

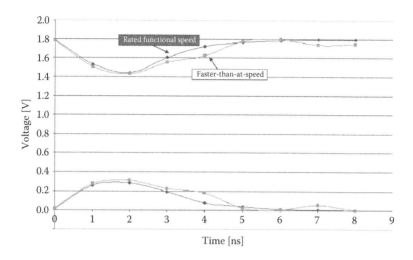

FIGURE 4.6
IR drop effects for rated functional speed f and faster-than-at-speed f' of a transition delay fault test application.

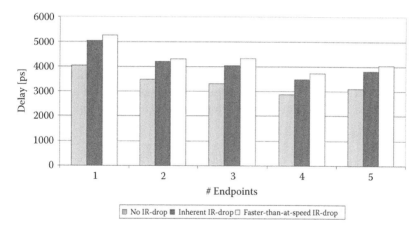

FIGURE 4.7
Path delay observed for five different endpoints, after applying a transition delay pattern, in three cases: (1) no IR drop; (2) inherent IR drop; and (3) faster-than-at-speed IR drop.

Figure 4.7 shows the path delay observed for five different endpoints in pattern P1 for three different cases: case 1, no IR drop effects; case 2, inherent IR drop at functional frequency; and case 3, faster-than-at-speed IR drop. The performance degradation due to faster-than-at-speed IR drop increased by up to 33% and 11% compared to no IR drop effect and inherent IR drop, respectively. The percentage increase in path delay across endpoints observing transitions in pattern P1 and comparison of case 3 with case 2 and case 1 is shown in Figure 4.8. For another pattern (P2)

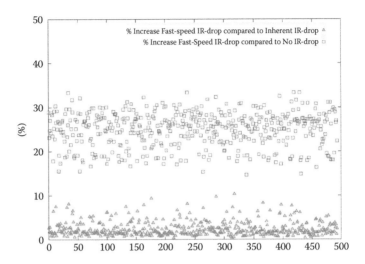

FIGURE 4.8
Percentage increase in path delay for faster-than-at-speed IR drop compared to inherent IR drop and no IR drop across endpoints in pattern P1.

during this case study, the maximum pattern delay for faster-than-at-speed application increased from 3,970 ps (case 1) to 6,050 ps (case 3), which was slightly beyond the selected fast-than-at-speed clock period ($T0 = 6,000$ ps) and resulted in a timing failure. Similarly, the percentage increase in path delay for pattern P2 is shown in Figure 4.9; observe that it shows higher performance degradation due to faster-than-at-speed IR drop compared to

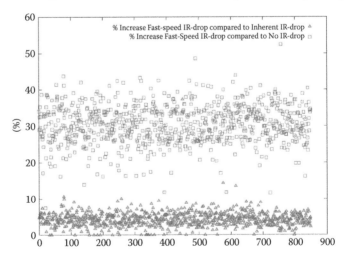

FIGURE 4.9
Percentage increase in path delay for faster-than-at-speed IR drop compared to inherent IR drop and no IR drop across endpoints in pattern P2.

pattern P1. Also, pattern P2 has more endpoints observing transitions than P1. This pattern would fail on testing due to IR drop effects rather than a small-delay defect. Therefore, it is important to consider the IR drop effects during faster-than-at-speed application along with the positive slack for detecting small-delay defects.

4.4 IR Drop Aware Faster-than-at-Speed Test Technique

As explained in the previous section, to improve the small-delay defect screening capability of a test pattern, the timing can be adjusted such that the path delay affected by a pattern is relatively close to the clock period (*pattern slack* near zero). However, any such timing adjustment (i.e., test pattern application at a higher test frequency) results in an increase in IR drop, and related performance degradation needs to be considered. Therefore, the problem is to determine the maximum test frequency at which a given transition fault test pattern can be applied without failing a good chip.

Each transition fault test pattern has varying delay distribution and switching activity, and it is difficult to perform detailed path delay analysis of each pattern as explained in Section 4.3. Also, even if such a detailed analysis were feasible, it may be difficult to apply each pattern at a different frequency. This might be either due to hardware limitations of the ATE to generate multiple higher frequencies or due to the test time limitation because of long synchronization times required for on-chip clock generators (PLLs). Therefore, to simplify the problem, the test pattern set is grouped into a user-defined number of subsets ($G1$, $G2$, ... , Gm, where m is the number of groups) with close pattern delay distribution and then determine the maximum frequency for each group considering faster-than-at-speed IR drop effects.

4.4.1 Pattern Grouping

To group the patterns with relatively close path delay distribution, the patterns were sorted in increasing order of pattern slack. The pattern slack is referred to as the least slack (maximum path delay) across all the endpoints in the respective pattern. The LOC transition fault pattern set (1,024 patterns) was generated using ATPG [Synopsys 2007b]. Figure 4.10 shows the maximum path delay affected in each pattern. Note that each pattern can affect various endpoints, but only the endpoint with the maximum delay (minimum slack) for each pattern is considered. The functional operating cycle time period is represented by $T = 10$ ns. Note that the affected paths in each of the patterns has a considerable amount of slack for the small-delay defects to escape during manufacturing test. This is because, in most cases, the ATPG tool found

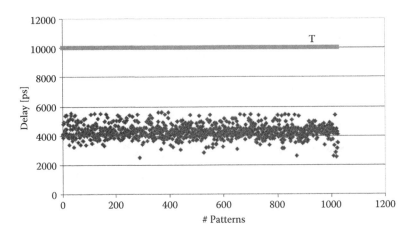

FIGURE 4.10
Maximum path delay affected in each pattern of the transition fault pattern set.

a shorter path to activate and propagate the transition fault effect, although this may not necessarily be the case for different designs.

Figure 4.11 shows the maximum path delay for each pattern in the resulting sorted pattern set. For faster-than-at-speed pattern application, the patterns with very close pattern slack distribution are grouped together (as shown in Figure 4.11). There are five groups ($m = 5$); $G1$ through $G5$, and the pattern slack range for each group was selected as 500 ps. A tighter pattern slack range can be selected (i.e., $m > 5$); it increases the number of groups and complexity as more processing will be required to perform IR drop analysis for each group. Assuming a fixed faster-than-at-speed cycle time for each group, the first pattern in each group will have the least pattern slack; hence,

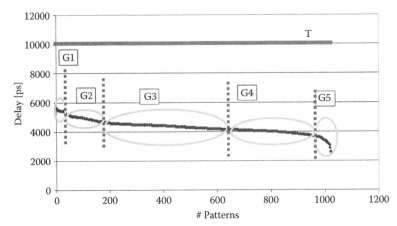

FIGURE 4.11
Sorted transition fault pattern set based on maximum path delay affected by each pattern.

it is used in determining the faster-than-at-speed test clock period for the respective group. If T'_{Gi} represents the new faster-than-at-speed clock period for group Gi, then T'_{Gi} can be formulated as $T'_{Gi} \geq T_{Gi} + |\Delta T'_{Gi}| + \tau_{su}$, where T_{Gi} represents the maximum path delay for pattern group Gi without considering IR drop effects, $\Delta T'_{Gi}$ represents the worst-case performance degradation due to faster-than-at-speed IR drop for group Gi, and τ_{su} is the setup time for the scan flip-flops.

4.4.2 Estimation of Performance Degradation $\Delta T'_{Gi}$

To determine the minimum test clock period T'_{Gi} for each pattern group, it is required to estimate the worst-case performance degradation $\Delta T'_{Gi}$ due to IR drop effects at the respective test frequency. Therefore, it is an iterative process to select a new faster-than-at-speed clock period T'_{Gi} and measure $\Delta T'_{Gi}$ for each pattern in the group until the performance degradation fails a pattern in the group. This is computationally expensive. To reduce the complexity, two patterns, P_{lsi} and P_{hsi}, are selected in each group Gi with the least slack and the highest SCAP [Ahmed 2007] during the launch-to-capture window, respectively. To further expedite the process, the SCAP model is used to measure the IR drop of P_{lsi} and P_{hsi} for each group. SCAP is an average power model; it provides a good measure of the IR drop as it considers both the simultaneous switching activity and the time frame in which it occurs. SCAP was originally proposed [Ahmed 2007] to address the issue of IR drop and its impact on path delay during TDF testing. SCAP can take into account both the number of transitions and the switching time frame window. The work of Ahmed and coworkers has more details [Ahmed 2007].

It is expected that pattern P_{hsi} will experience the highest IR drop (ΔV) (i.e., SCAP) in the group for any applied test frequency. To measure the worst performance degradation, the IR drop for pattern P_{hsi} is then measured at T_{Gi}, that is, the minimum possible test clock period without considering IR drop effects. The corresponding worst cell delay degradation $k\_volt*\Delta V$ if applied to pattern P_{lsi} with the maximum path delay in the group Gi would provide the worst possible path delay T'_{Gi} that can be used as the fastest at-speed clock period for the respective group. This procedure will provide enough timing margin for the entire pattern group considering faster-than-at-speed IR drop effects.

The new faster-than-at-speed test pattern application framework is shown in Figure 4.12. The framework assumes that a delay pattern set is available, generated using an ATPG tool. Two programming language interface (PLI) routines, *monitorDelay* and *calculateSCAP*, have been developed. The PLI provides a standard interface to the internal data, such as the nets switching inside the design during simulation. The PLIs can be plugged into any gate-level simulator (here Synopsys VCS [Synopsys 2007b]). *monitorDelay* monitors the switching activity across the functional pin (D) of each scan flip-flop and creates a pattern delay profile with the entire path delay

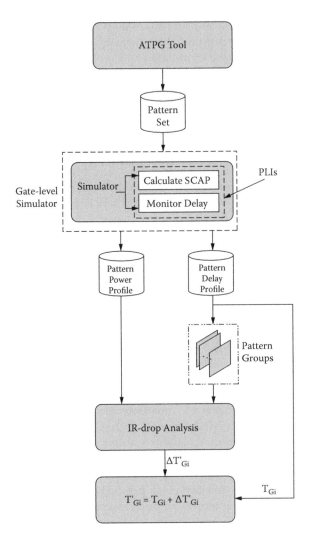

FIGURE 4.12
New faster-than-at-speed test pattern application framework.

distribution of each pattern. The maximum path delay of each pattern is then extracted, and the patterns are grouped into subsets with relatively close path delay distribution. *calculateSCAP* creates the pattern power profile and measures the SCAP of each pattern in the VDD and VSS networks. The capacitance of each gate instance is extracted from the RC parasitics file (standard parasitics exchange format, SPEF) generated with an extraction tool (Synopsys STAR-RCXT [Synopsys 2007b]). It then uses the switching activity and parasitics information to measure the SCAP value of pattern P_j:

$$SCAP_j = \Sigma i(Ci * VDD^2)/STW_j$$

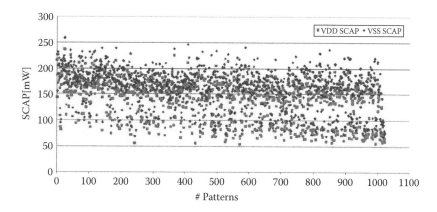

FIGURE 4.13
Switching cycle average power (SCAP) in VDD and VSS networks for each transition fault test pattern.

where C_i is gate ith load capacitance, and STW_j is the switching time frame window of pattern P_j. SCAP is calculated for both VDD and VSS networks for each pattern during the launch-to-capture window. STW is the time span during which the entire switching activity occurs between the two at-speed capture pulses. Figure 4.13 shows the SCAP measured in VDD and VSS networks in each transition fault test pattern during the launch-to-capture window.

Using the PLI to calculate SCAP avoids large industry standard VCD file generation for estimation of switching power. However, the VCD file is still required for IR drop analysis, but it is performed only on one pattern in each group with the highest SCAP value, hence reducing the CPU run time significantly. Figure 4.14 shows (VDD + VSS) SCAP in each pattern for the reordered pattern set based on maximum path delay. It can be observed that the average SCAP increased from group $G1$ through $G5$ as the switching time frame window and the maximum path delay decreased.

After determining the pattern in each group with the highest SCAP value (P_{hsi} of group Gi circled in Figure 4.14), the entire net toggle activity for this particular pattern is captured in a VCD file, and the worst-case IR drop analysis is performed. In the next step, the corresponding worst performance degradation $\Delta T'_{Gi}$ is estimated, and the faster-than-at-speed test period for each group is determined as explained in Section 4.4.2.

Note that, in the proposed method, the patterns were sorted based on the longest path they affected. However, each pattern can affect many paths. Reducing the slack based on the longest path affected for each pattern ensures detection of small-delay defects on most paths sensitized by the pattern but may not on all of them. Therefore, if a pattern exists that affects a long path in addition to a short path, then the method can put the pattern into two different groups based on the length of the paths. In other words, the same pattern can be applied to the chip twice but at different frequencies.

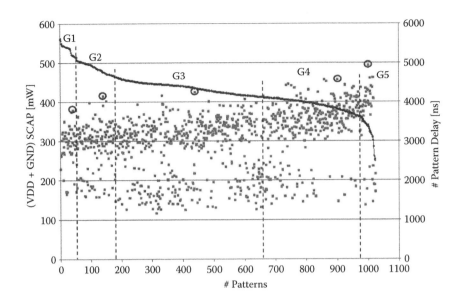

FIGURE 4.14
(VDD + VSS) SCAP for reordered pattern set based on maximum path delay.

The method can also mask the output of the longer path when testing the shorter one. This method can be easily employed in current DFT flows and work with existing commercial compression tools as no change is required to the ATPG.

4.5 Experimental Results

The LOC transition fault pattern set (1,024 patterns) was generated using ATPG [Synopsys 2007b]. As explained in the previous section, the framework divides the entire pattern set into multiple groups. The groups are divided based on the relative closeness of pattern slack. Here, the pattern slack range was selected as 500 ps to generate a reasonable number of pattern groups ($m = 5$ groups in this case). However, this is configurable, and decreasing the pattern slack range increases the number of pattern groups.

Table 4.1 shows the results obtained for each group $G1$ through $G5$ after sorting the pattern set based on maximum path delay of each pattern. The second column in the table shows the number of patterns in each group. The worst delay T_{Gi} of a group is the maximum delay of a pattern in each group (column 3) without considering the IR drop effect. To estimate the worst-case average IR drop in each group, a pattern with the highest SCAP (P_{hsi}) value was selected, and an IR drop analysis was performed for a clock

TABLE 4.1

Pattern Grouping and Worst IR Drop Results per Pattern Group

Group	Number of Patterns	Worst Delay T_{Gi}	Worst Average ($\Delta VDD_{IR\,drop}$)	Worst Average ($\Delta VSS_{IR\,drop}$)	Worst Average ($\Delta(VDD + VSS)_{IR\,drop}$)
G1	54	5,618	0.167	0.148	0.315
G2	135	5,072	0.183	0.16	0.343
G3	468	4,618	0.235	0.213	0.448
G4	321	4,118	0.251	0.225	0.476
G5	46	3,604	0.244	0.239	0.483

period T_{Gi} as explained in Section 4.3. Columns 4 and 5 in Table 4.1 show the worst-case IR drop for VDD ($\Delta VDD_{IR\,drop}$) and VSS ($\Delta VSS_{IR\,drop}$), respectively. Note that, as the switching time frame window decreased from group G1 through G5, the average (VDD + VSS) IR drop increased (column 6). The IR drop depends on both the switching activity and time frame window (as defined in SCAP) in each pattern, and Figure 4.14 shows that not all patterns in group G5 will have a high IR drop. Also, the highest SCAP pattern (P_{hsi}) in group G4 observed a higher VDD IR drop compared to G5, but the effective (VDD + VSS) IR drop was less than G5.

To take into account the performance degradation effect due to faster-than-at-speed IR drop, the design timing information at the respective effective voltage ($\Delta V = \Delta VDD_{IR\,drop} + \Delta VSS_{IR\,drop}$) needs to be generated for each group. For example, for group G1, the effective voltage is $VDD - \Delta V = 1.8 - (0.167 + 0.148) = 1.485$ V. The design timing information at the new operating voltage condition can be generated by two methods.

First, the standard cell library can be characterized at the new operating voltage, and a timing analysis tool can be used to generate the SDF file. The second method is to take the effective voltage reduction and apply cell delay degradation [$Scaled\_Cell\_Delay = Cell\_Delay * (1 + k\_volt * \Delta V)$] to generate the new design timing information. Hence, for group G1 with effective voltage reduction of $\Delta V = 0.315$ V due to IR drop, each cell delay on the longest path will increase by 29% at the faster-than-at-speed test period T'_{Gi} compared to no IR drop at the functional clock period.

Our technique applies the voltage drop caused by the pattern with highest SCAP to the worst-case slack pattern. Here, it is assumed that the worst-slack path will experience the worst voltage drop. The pattern with the maximum delay in the group is then simulated with the delay scaling factor to obtain the worst performance degradation due to faster-than-at-speed IR drop, which is shown in Column 3 of Table 4.2. The delay scaling factor is applied to both cell and interconnect delays in the design. Note that the percentage of worst performance degradation (column 3) increases from G1 to G5 as the effective (VDD + VSS) voltage drop increases. The worst performance degradation measured by simulation for group G1 was 31.2%,

TABLE 4.2

Estimated Faster-than-at-Speed Test Clock Results for Different Pattern Groups

Group	Worst Delay (T_{Gi})	Worst Performance Degradation ΔT_{Gi} (ps)	Faster-than-at-Speed Period ($T' = T_{Gi} + \Delta T_{Gi} + \tau_{su}$)
G1	5,618	1,752 (31.2%)	7,802
G2	5,072	1,718 (33.8)	7,114
G3	4,618	2,047 (44.3%)	6,638
G4	4,118	1,940 (47.1%)	6,082
G5	3,604	1,722 (47.8%)	5,364

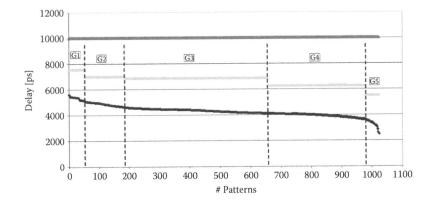

FIGURE 4.15
Transition fault pattern groups with their respective faster-than-at-speed clock periods.

which is slightly greater than 29%. This is because scaling was applied to cell and interconnect delay on both clock and data paths, and there was a difference in clock insertion delay between source and sink flops.

Finally, the resulting worst-case performance degradation was used to determine the faster-than-at-speed clock period ($T'_{Gi} = T_{Gi} + \Delta T_{Gi} + \tau_{su}$) for the respective pattern group (see column 4 in Table 4.2). Figure 4.15 shows the original rated functional period and faster-than-at-speed clock timing for each of the groups considering IR drop effects. Note that there is extra slack ΔT_{Gi} provided by the new technique between the maximum delay of the pattern group and the faster-than-at-speed clock period to take into account the performance degradation due to an increase in IR drop effects.

4.6 Conclusions

This chapter presented a detailed analysis of a new faster-than-at-speed technique utilized for small-delay fault detection. The analysis illustrated that the IR drop is exacerbated during faster-than-at-speed pattern application (up to 15% compared to the IR drop at the rated functional speed), and it is important to consider the performance degradation of the design due to an increase in IR drop effects. A new framework for applying transition fault test patterns at a faster-than-at-speed rate was proposed considering both the performance degradation due to adverse IR drop effects and positive slack. The proposed technique grouped the pattern sets based on their affected maximum delay and determined the worst-case performance degradation for each pattern group. This ensures avoiding false identification of good chips as faulty due to IR drop effects rather than small-delay defects.

4.7 Acknowledgment

The work of Mohammad Tehranipoor was supported in part by Semiconductor Research Corporation (SRC) under Contract 1587 and by the National Science Foundation (NSF) under grants ECCS-0823992 and CCF-0811632.

References

[Ahmed 2005] N. Ahmed, C. Ravikumar, M. Tehranipoor, and J. Plusquellic, *At-Speed Transition Fault Testing with Low Speed Scan Enable*, IEEE Computer Society, 2005.

[Ahmed 2006b] N. Ahmed, M. Tehranipoor, and V. Jayram, Timing-based delay test for screening small delay defects, in *Proceedings IEEE Design Automation Conference*, 2006.

[Ahmed 2007] N. Ahmed, M. Tehranipoor, and V. Jayaram, Supply voltage noise aware ATPG for transition delay faults, in *Proceedings IEEE VLSI Test Symposium*, 2007.

[Ahmed 2010] N. Ahmed and M. Tehranipoor, A novel faster-than-at-speed transition delay test method considering IR-drop effects, in *IEEE Transactions on CAD*, 2010.

[Allampally 2005] S. Allampally, V. Prasanth, R. K. Tiwari, and K. S., Small delay defect testing in DSM technologies—Needs and considerations, Synopsys Users Group Conference, 2005.

[Benware 2003] B. Benware, C. Schuermyer, N. Tamarapalli, K.-H. Tsai, S. Ranganathan, R. Madge, J. Rajski, and P. Krishnamurthy, Impact of multiple-detect test patterns on product quality, in *ITC 2003 Test Conference Proceedings*, 1, pp. 1031–1040, 2003.

[Cadence 2004] Cadence Inc., *User Manual for Cadence Encounter Toolset Version 2004.10*, Cadence Inc., San Jose, CA, 2004.

[Cadence 2005] Cadence Inc., *0.18μm Standard Cell GSCLib Library Version 2.0*, Cadence Inc., San Jose, CA, 2005.

[Cadence 2006] Cadence Inc., *User Manual for Encounter True-time Test ATPG*, Cadence Inc., San Jose, CA, 2006.

[Foster 1976] R. Foster, Why consider screening, burn-in, and 100-percent testing for commercial devices? *IEEE Transactions on Manufacturing Technology* 5(3), pp. 52–58, September 1976.

[Gaisler 2006] Gaisler Research, 2006. Available http://gaisler.com.

[Gupta 2004] P. Gupta and M. Hsiao, ALAPTF: A new transition fault model and the ATPG algorithm, in *Proceedings IEEE International Test Conference*, 2004, pp. 1053–1060.

[Hao 1993] H. Hao and E. McCluskey, Very-low-voltage testing for weak CMOS logic ICs, in *Proceedings IEEE International Test Conference*, 1993, pp. 275–284.

[Kruseman 2004] B. Kruseman, A.K. Majhi, G. Gronthoud, and S. Eichenberger, On hazard-free patterns for fine-delay fault testing, in *Proceedings IEEE International Test Conference*, 2004.

[Lee 2005] B.N. Lee, L.-C. Wang, and M. Abadir, Reducing pattern delay variations for screening frequency dependent defects, in *Proceedings IEEE VLSI Test Symposium*, 2005, pp. 153–160.

[Lin 2003] X. Lin, R. Press, J. Rajski, P. Reuter, T. Rinderknecht, B. Swanson, and N. Tamarapalli, High frequency, at-speed scan testing, *IEEE Design and Test Computers* 20(5), pp. 17–25, September/October 2003.

[Lin 2006] X. Lin, K.-H. Tsai, C. Wang, M. Kassab, J. Rajski, T. Kobayashi, R. Klingenberg, Y. Sato, S. Hamada, and T. Aikyo, Timing-aware ATPG for high quality at-speed testing of small delay defects, in *Proceedings IEEE Asian Test Symposium*, 2006, pp. 139–146.

[Qiu 2004] W. Qiu, J. Wang, D. Walker, D. Reddy, X. Lu, Z. Li, W. Shi, and H. Balachandran, *K* longest paths per gate [KLPG] test generation for scan-based sequential circuits, in *Proceedings IEEE International Test Conference*, 2004, pp. 223–231.

[Saxena 2002] J. Saxena, K. Butler, J. Gatt, R. Raghuraman, S. Kumar, S. Basu, D. Campbell, and J. Berech, Scan-based transition fault testing—implementation and low cost test challenges, in *Proceedings IEEE International Test Conference*, 2002, pp. 1120–1129.

[Synopsys 2007b] Synopsys Inc., *User Manual for Synopsys Toolset Version 2007.03*, Synopsys Inc., Mountain View, CA, 2007.

5

Circuit Path Grading Considering Layout, Process Variations, and Cross Talk

Ke Peng, Mahmut Yilmaz, and Mohammad Tehranipoor

CONTENTS

5.1 Introduction

The semiconductor industry has come to rely heavily on delay testing for high-defect coverage of manufactured digital systems. This is because testing delay defects (also referred as timing-related defects) has become extremely vital for ensuring product quality and in-field reliability in the very deep-submicron (VDSM) regime. In general, these defects refer to any type of physical and nonphysical defects, which are introduced either by imperfect manufacturing processes or by variations in design parameters. Such defects can cause the chips to fail by introducing extra signal propagation delay to produce an invalid response when the chip operates at its operating frequency. A small-delay defect (SDD) is one kind of such timing defects. SDDs were not seriously considered when testing designs at higher technology nodes because of their small size relative to the timing margins allowed by the maximum operating frequency of a design. Although the delay introduced by each SDD is small, the overall impact can be signifi-cant if the sensitized path is long or critical. As the shrinking of technology geometries and increasing of operating frequency of the design continue, the available timing slack becomes smaller. Therefore, SDDs have a good chance to add enough additional delay to a path to adversely impact the circuit timing and make that part deviate from its functional specifications. Studies have shown that a large portion of failures in delay-defective parts are due to SDDs for the latest technologies [Mattiuzzo 2009; Ahmed 2006b]. Therefore, SDDs require serious consideration to increase the defect cover-age and test quality or decrease the number of test escapes (i.e., increase in-field reliability), denoted by defective parts per million (DPPM).

5.1.1 Commercial Methodologies for SDD Detection

The transition delay fault (TDF) and path delay fault are two prevalent fault models widely used in industry. The TDF model assumes that there is a *slow-to-rise* or *slow-to-fall* fault at each gate output in the design [Savir 1992, 1994b]. A delay fault at a gate output, when it is sufficiently large (gross-delay defect), can cause a logic failure when the signal is propagated through a path to the test observation points. The path delay fault model assumes a cumulative delay through all the gates and interconnects on a predefined path. The path delay fault is superior to the TDF in modeling capacity by addressing distributed defects that affect the entire path. The major draw-back of the path delay fault model is that it requires a large test pattern set, and it may also be impossible to excite all the paths in the design because the number of paths in the design increases exponentially with circuit size. Thus, path delay fault models are only used for selected long/critical paths in the design [Goel 2009]. In practice, most timing defects are targeted by the TDF model and detected by test patterns generated from TDF automatic test pattern generation (ATPG) tools.

TDF ATPG targets delay defects by generating vector $v1$ to initialize the circuit nodes. Then, it generates another vector, $v2$, to launch a slow-to-rise or slow-to-fall transition through the target fault site, ensuring the transition can be propagated to the endpoint of the path, which can be a primary out-put or pseudoprimary output (scan flip-flop). If the signal does not propagate to the endpoint at the capture time, the path for signal propagation must be longer than the specific cycle time, and the incorrect data are captured that denote the delay defect through the sensitized path. The challenge, how-ever, is to target SDDs by the TDF model and TDF-based pattern generation procedure in an effective manner.

Experiments have demonstrated that a TDF test pattern set can achieve a defect coverage level that stuck-at patterns alone cannot, and it can also detect some of the SDDs. Unfortunately, such pattern sets have shown limited ability for detecting SDDs in the device and meeting the high SDD test coverage requirements in industry, which is very low or close-to-zero DPPM for some critical applications, such automotive or medical systems. The traditional ATPG tools were developed to target gross delay defects with minimum run time and pattern count, rather than targeting SDDs. To minimize run time and pattern count, TDF ATPGs were developed to be timing unaware and activate TDFs via the easiest sensitization and propagation paths, which are often shorter paths [Ahmed 2006b]. Note that a TDF can only be detected if it causes a signal transition that exceeds the slack of the affected path. Considering this fact, an SDD may escape the traditional TDF testing and may cause failure in the field if a long path passes through it [Ahmed 2006b; Majhi 2000]. Therefore, it is necessary to detect SDDs through long paths.

Commercial timing-aware ATPG tools (e.g., latest versions of Synopsys TetraMAX [Synopsys 2007a] and Mentor Graphics FastScan [Mentor 2006]) have been developed to deal with the deficiencies of traditional timing-unaware ATPGs. Similar to path delay fault ATPGs, timing-aware ATPG targets each undetected fault along the path with minimum timing slack, or a small timing slack according to the user specification, which is the long path running through the fault site. If a fault is detected through the minimum slack path, it is categorized as detected by simulation (DS) and removed from the fault list. The generated pattern will be filled either randomly or using a compression engine. Then, fault simulation will be run to identify the detected SDDs as well as those additional fortuitously detected SDDs that might have been detected along their least-slack paths. If a fault is detected along a path with a slack larger than the specified threshold (least slack), it is identified as partially detected. Such faults will continue to be targeted every time ATPG generates a new pattern; the fault will be dropped from the fault list if, at some point during the pattern generation process, it is detected through a path that meets the slack requirement.

The main drawback of the timing-aware ATPG algorithm is that it wastes a lot of time operating on faults that do not contribute to SDD coverage and results in a large number of patterns. Experimental results have demonstrated that timing-aware ATPGs will result in a significantly long CPU run time and pattern count. Furthermore, they seem (1) ineffective in sensitizing a large number of long paths and (2) incapable of taking into account important design parameters like process variations, cross talk, and power supply noise, which are important sources for inducing small delay in VDSM designs [Yilmaz 2008c,2008b].

Due to its underlying algorithms, the *n-detect* ATPG can also be an effective method for SDD detection, even without timing information for the design. For each target fault, the *n-detect* ATPG will generate patterns trying to detect it *n* times, through different paths [Amyeen 2004]. Therefore, if there is a

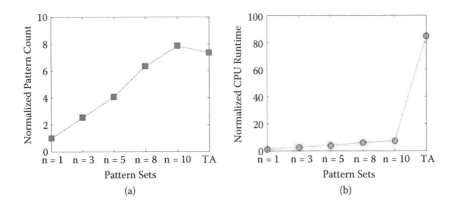

FIGURE 5.1
Normalized CPU run time (a) and pattern count (b) of different pattern sets for usb_funct benchmark circuit.

sensitizable long path running through the target fault site with a large value of n for n-*detect* ATPG, the ATPG tool will have a good chance of detecting faults via their possible long paths. In other words, n-*detect* ATPG can result in high-quality patterns for screening SDDs. Furthermore, experiments have demonstrated that the n-*detect* ATPG requires much less CPU run time when compared with timing-aware ATPG. However, the significantly large pattern count for large n limits the usage of n-*detect* ATPG in practice.

Figure 5.1 presents the normalized CPU run time and pattern count for 1-*detect*, n-*detect* (n = 3, 5, 8, 10), and timing-aware TDF ATPGs for the International Workshop for Logic Synthesis (IWLS) benchmark usb_funct [Gupta 2004]. The CPU run time and pattern count are normalized with respect to 1-*detect* ATPG. It can be seen from the figure that as n increases, the CPU run time of n-*detect* timing-unaware ATPG increases. Timing-aware ATPG consumes a much longer CPU run time compared with timing-unaware ATPGs (over 80 times that of 1-*detect* ATPG). The pattern count of n-*detect* ATPG increases almost linearly with n for this benchmark. The timing-aware ATPG results in a pattern count similar to 10-*detect* ATPG for this benchmark. In conclusion, compared with 1-*detect* timing-unaware ATPG, timing-aware ATPG can generate high-quality SDD patterns with the penalty of a large pattern count and significant CPU run time, and n-*detect* ATPG will result in high-quality SDD patterns with a large pattern count and less CPU run time increase.

Another way of detecting SDDs is simply to perform faster-than-at-speed testing. This kind of technique increases the test frequency to reduce the positive slack of the path to detect SDDs on target paths. However, the application of this method is limited since (1) the on-chip clock-generation circuits must be overdesigned to meet the requirements of the faster-than-at-speed testing to provide various high-frequency steps and frequency sweep ranges, which would make it expensive and difficult to generate given the process variations and (2) the methodology may result in false identification of good chips as faulty due to the reduced cycle time and increased IR drop [Ahmed 2006a], leading to unnecessary yield loss.

5.1.2 Academic Proposals for SDD Detection

Several techniques have been presented in the past targeting the increase in test quality for screening SDDs.

- An as-late-as-possible transition fault (ALAPTF) model was proposed to launch one or more transition faults at the fault site as late as possible to accumulate the SDDs along its way and detect these faults through the least-slack path [Gupta 2004]. This method requires long CPU run time compared to traditional ATPGs.

- Qiu and coworkers [Qiu 2004] proposed a method to generate K longest paths per gate for testing transition faults. However, the longest path through a gate may also be a short path for the design and thus may not be efficient for SDD detection. This method also suffers from high complexity, long CPU run time, and large pattern count.

- Some authors [Majhi 2000] proposed a delay fault coverage metric to detect the longest sensitized path affecting a TDF site. It is based on a robust path delay test and attempts to find the longest sensitizable path passing through the target fault site and generating a slow-to-rise or slow-to-fall transition. It is impossible to implement this method on large industry circuits since the number of long paths increases exponentially with the circuit size.

- Others [Goel 2009] proposed two hybrid methods using *1-detect* and timing-aware ATPGs to detect SDDs with reduced pattern count. These methods first identify a subset of transition faults that are critical and should be targeted by the timing-aware ATPG. Then, top-off ATPG is run on the undetected faults after timing-aware ATPG to meet the fault coverage requirement. The efficiency of this method is questioned since it still results in a pattern count much larger than traditional *1-detect* ATPG.

- A method based on static timing analysis was proposed [Ahmed 2006b] to generate and select patterns that sensitize long paths. It finds long paths, intermediate paths, and short paths to each observation point using static timing analysis tools. Then, intermediate path and short path observation points are masked in the pattern generation procedure to force the ATPG tool to generate patterns for long paths. Next, a pattern selection procedure is applied to ensure pattern quality.

- An output-deviation-based method was proposed [Yilmaz 2008c]. This method defines gate delay defect probabilities (DDPs) to model delay variations in a design. Gaussian distribution gate delay is assumed, and a delay defect probability matrix (DDPM) is assigned to each gate. Then, the signal transition probabilities are propagated to outputs to obtain the output deviations, which are used for pattern evaluation and selection. However, in the case of a large number of gates along the paths, with this method, the calculated output deviation metric can saturate, and similar output deviations (close to 1) can be obtained for both long and intermediate paths (relative to clock cycle). Since in today's modern designs there exist a large number of paths with large depth in terms of gate count, the method based on output deviation may not be effective. A similar method was developed [Yilmaz 2008b] to take into account the interconnect contribution to the total delay of sensitized paths. Unfortunately, this work also suffers from a similar problem.

- A false-path-aware statistical timing analysis framework was proposed [Lion 2002]. It selects all logically sensitizable long paths using worst-case statistical timing information and obtains the true timing information of the selected paths. After obtaining the critical long paths, it uses path delay fault test patterns to target them. This method will be limited by the same constraints that the path delay fault test experiences.

- Lee et al. [Lee 2006] proposed path-based and cone-based metrics for estimating path delay under test. These methods are not accurate due to their dependence on gate delay models; unit gate delay and differential gate delay models, which were determined by the gate type, number of fan-ins and fan-outs; and the transition type at the output. Furthermore, this method is also based on static timing analysis. It does not take into account pattern information, pattern-induced noise, process variations, and their impact on path delays.

In general, the most effective way of detecting an SDD is to detect it via long paths. The commercial timing-aware ATPG tools and most previously proposed methods for detecting SDDs have relied on standard delay format (SDF) files generated during physical design flow [Synopsys 2007a; Mentor 2006]. However, the path length is greatly impacted by process variations, cross talk, and power supply noise. A short path may become long because of one or a combination of such effects. Some of these effects may also make a long path become short. For instance, in certain circumstance, crosstalk effects can speed up the signal propagation and shorten the path length. SDF files are pattern independent and are incapable of taking these effects into consideration. Furthermore, the complexity of today's ICs and shrinking process technologies have made the design features more probabilistic. Thus, it is necessary to perform statistical timing analysis before evaluating the path length. Both traditional and timing-aware ATPGs are not capable of addressing these statistical features. In this chapter, a novel pattern evaluation and selection procedure is presented that results in high-quality tests for screening SDDs. From the implementation of the proposed procedure on several benchmarks, this procedure can result in a pattern count as low as a traditional *1-detect* pattern set and long-path sensitization and SDD detection similar to or even better than the *N-detect* or timing-aware pattern set. This procedure is capable of adding important design parameters, such as process variations, cross talk, power supply noise, and the like. In this chapter, only process variations and cross talk are added in the pattern evaluation and selection procedure.

5.2 Analyzing SDDs Induced by Variations

5.2.1 Impact of Process Variations on Path Delay

In reality, the parameters of fabricated transistors are not exactly the same as that in the simulated design. In fact, the parameters are different from die to die, wafer to wafer, and lot to lot. These variations, including impurity concentration densities, oxide thicknesses, and diffusion depths caused by nonuniform conditions during the deposition or the diffusion of the impurities, will directly result in deviations in transistor parameters, such as threshold voltage, oxide thickness, W/L ratios, as well as variations in the widths of interconnect wires [Rabaey 2003]. These process variations will have an impact on design performance (increase or decrease gate and interconnect delays) to a large extent in the latest technologies. Thus, it is necessary to take these effects into consideration in the path/pattern evaluation procedure.

Due to the impact of process variations, the delay of each path segment (gate and interconnect) is assumed to be a random variable X with mean value μ and standard deviation σ, rather than a fixed value. In these experiments, Monte Carlo simulations were run using a 180-nm Cadence Generic Standard Cell Library to obtain the delay distributions (mean value μ and standard deviation σ) for all gates in the library. For each gate, Monte Carlo simulations were run with

1. Different input switching combinations
2. Different output load capacitances
3. Process variation parameters:
   a. Transistor gate length L: $3\sigma = 15\%$
   b. Transistor gate width W: $3\sigma = 15\%$
   c. Threshold voltage V_{th}: $3\sigma = 15\%$
   d. Gate oxide thickness t_{ox}: $3\sigma = 3\%$

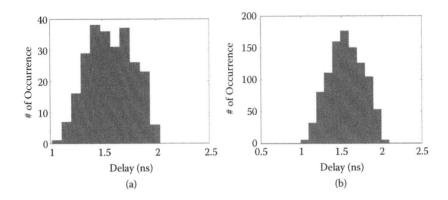

FIGURE 5.2
Delay histograms obtained from Monte Carlo simulation results after (a) 250 and (b) 1,000 simulation runs.

Figure 5.2 shows an example of the Monte Carlo simulation results on a NAND3X1 gate, with the above variations, input switching from {011} to {111}, and output load capacitance 1.0 pf. It is clear that the mean value μ and standard deviation σ for the random variable corresponding to the gate delay are bounded in a certain range (e.g., 1 to 2.5 ns for this example), and as the number of Monte Carlo simulations increases, the probability density function (PDF) of the gate becomes closer to a Gaussian distribution. With a large number of Monte Carlo simulation runs, more accurate PDFs can be obtained, and more CPU run time is needed. To trade off the accuracy and CPU run time, 250 Monte Carlo simulations were run for each input-output combination of a gate. In this way, the PDF of each standard gate in the library could be obtained, and a gate-PDF database considering process variations could be created.

The interconnect PDFs can be obtained by calculating their lengths. In this experiment, the interconnect length was extracted from the layout. Different variations between metal layers and vias were taken into consideration when calculating the delay distributions. There are six available metal layers in the library. For a metal segment with unit delay, 3σ variations were 30%, 20%, 15%, 10%, 5%, and 5% for metal 1 to metal 6, respectively. The 3σ variations for vias in metal layers 1 through 5 were 50% [ITRS 2008].

In this chapter, it is assumed that delay variations on path segments (gates or interconnects) were independent from each other. It was seen that the mean and variance of the delay of each path segment were bounded. Therefore, Lindeberg's condition is satisfied, and the central limit theorem (CLT) holds [Koralov 2007]. The CLT states that the sum of a sufficiently large number of independent random variables, each with finite mean and variance, approaches a normal distribution. Let the random variable X_p

denote the path delay. According to the CLT, the random variable X_p can be calculated using Equation 6.1.

$$X_p = \sum_{i=1}^{n} X_{si} \tag{6.1}$$

where n is the number of segments along the path, and X_{si} is the delay of the ith segment along the path. The mean and standard deviation of X_p can be calculated using Equations 6.2 and 6.3, respectively.

$$\mu_p = \sum_{i=1}^{n} \mu_{si} \tag{6.2}$$

$$\sigma_p = \sqrt{\sum_{i=1}^{n} \sigma_{si}^2} \tag{6.3}$$

where μ_{si} and σ_{si} are the mean delay and standard deviation for segment i, respectively. In this way, the PDF of each target path can be calculated. Figure 5.3 presents an example of segment PDF propagation and path PDF calculation.

5.2.2 Impact of Cross Talk on Path Delay

There are millions of interconnect segments running in parallel in a design, with parasitic coupling capacitances between them, introducing crosstalk effects and impacting the circuit delay characteristics and performance. As the shrinking of technologies continues, the distance between interconnect wires is reduced, and the coupling capacitance between them is increased; thus, the crosstalk effects between them are increased. Therefore, it is necessary to

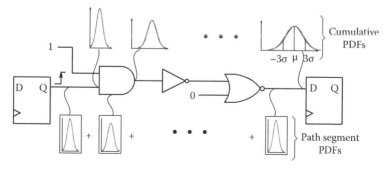

FIGURE 5.3
Gate and interconnect PDF propagation through the target path.

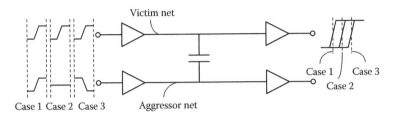

FIGURE 5.4
Impact of transition direction on victim propagation delay.

take crosstalk effects into consideration for accurate path/pattern evaluation and selection.

The crosstalk effects introduced by parasitic coupling capacitance between a target net (victim) and its neighboring nets (aggressors) may either speed up or slow the delays on both victim and aggressor nets, according to the transition direction, transition arrival time, as well as coupling capacitance between the victim and aggressor nets [Chen 1997]. Take the single-aggressor case as an example: If the aggressor and victim nets have the same transition direction, delay on the victim net will be decreased (known as a speed up). Otherwise, when aggressor has an opposite transition direction with respect to the victim net, delay on the victim net will be increased (known as a slow down) as shown in Figure 5.4. Furthermore, if the aggressor and victim net switch at the same time, the crosstalk effects are usually maximized.

Since transitions on aggressors and victim have different directions and arrival times, a set of Semiconductor Program with Integrated Circuit Emphasis (SPICE) simulations was performed to analyze the crosstalk impact on each other and set up a realistic crosstalk model for accurate path/pattern

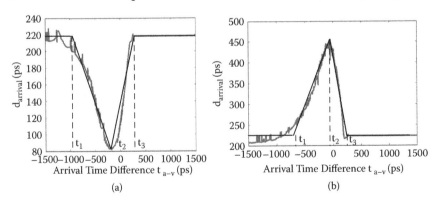

FIGURE 5.5
Impact of aggressor arrival time on victim propagation delay when victim and aggressor nets have (a) the same transition direction and (b) an opposite transition direction. Coupling capacitance, 0.1 pf.

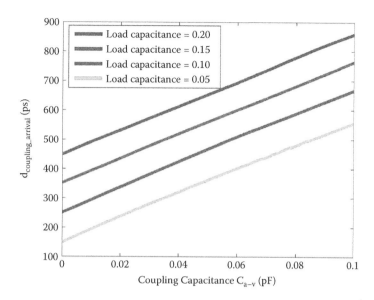

FIGURE 5.6

Impact of coupling capacitance on the victim net when the victim and aggressor nets have the same arrival times but opposite transition directions.

evaluation. Figure 5.5 presents the simulation results on crosstalk effects between two neighboring interconnects (one victim and one aggressor) with various arrival time differences and a fixed coupling capacitance. The times $t1$, $t2$, and $t3$ in the figure represent the break points of the curve fitting. The parameter t_{a-v} denotes the arrival time difference between transitions on aggressor and victim nets, and $d_{arrival}$ represents the victim net delay considering the impact of arrival time difference. It is seen that when the aggressor and victim nets have the same transition direction (see Figure 5.5a), the victim net will speed up. Otherwise, the victim net will slow down (see Figure 5.5b). Furthermore, the crosstalk effect on the victim net is maximized when the transition arrival time of aggressor and victim nets are almost same ($t_{a-v} \approx 0$).

Figure 5.6 presents the simulation results for crosstalk effects between two neighboring interconnects with various coupling and load capacitances. In this experiment, the signal arrival time difference between the aggressor and victim nets and signal transition directions were constant. In this figure, C_{a-v} is the coupling capacitance between the aggressor and victim nets, and $d_{coupling\_arrival}$ denotes the victim net delay considering the impact of coupling capacitance size. It is seen that for different load capacitances, the propagation delay on the victim net increased linearly with the coupling capacitance. For the same transition direction case, the crosstalk delay decreased linearly.

In the literature on statistical methods, it has been demonstrated that the least-squares technique is useful for fitting data to a curve [Recktenwald

2000]. Instead of solving the equations exactly, the least-squares technique tries to minimize the sum of the squares of the residuals. The least-squares curve fitting was applied to the simulation results shown in Figure 5.5 and approximated a piecewise function relationship between the interconnect delay and the aggressor-victim alignment as shown in Equation 6.4.

$$
d_{arrival} =
\begin{cases}
d_{orig} & 0 \leq t_{a-v} < t_1 \\
a_0 t_{a-v} + a_1 & t_1 \leq t_{a-v} < t_2 \\
b_0 t_{a-v} + b_1 & t_2 \leq t_{a-v} < t_3 \\
d_{orig} & t_3 \leq t_{a-v}
\end{cases}
\tag{6.4}
$$

where d_{orig} is the original interconnect delay without considering crosstalk effects. Here, t_{a-v} is the arrival time difference between aggressor and victim nets; a_0 and b_0 are the curve slopes between timing windows $[t_1, t_2]$ and $[t_2, t_3]$ (see Figure 5.5), respectively. a_1 is the arrival time delay of the victim net with the conditions $t_{a-v} = 0$ and $t_2 > 0$. Similarly, b_1 is the arrival time delay when $t_{a-v} = 0$ and $t_2 < 0$.

After considering the impact of the transition direction and arrival time, the impact of coupling capacitance was taken into account. Since the impact of coupling capacitance on the victim delay is approximately linear, it is much easier to do the least-squares curve fitting and set up the functional relationship, as shown in Equation 6.5.

$$
d_{coupling\_arrival} = a' d_{arrival} C_{a-v}
\tag{6.5}
$$

where $d_{coupling\_arrival}$ is the propagation delay considering the impact of the transition direction, arrival time, and coupling capacitance between the aggressor net and the target victim net. The factor a' is negative for the same transition direction and positive for the opposite transition direction case. C_{a-v} is the coupling capacitance between the aggressor and victim nets. Furthermore, these parameters are highly dependent on technology nodes. For different technologies, these parameters are different.

With this crosstalk model, the impact of a single aggressor on the victim delay can be calculated. For the case with multiple aggressors, the impacts of aggressors are applied one by one, according to their arrival time. Figure 5.7 shows the crosstalk impacts on interconnect PDFs. It was implemented under the assumption that crosstalk effects will only have an impact on the mean delay value of the victim net rather than delay variation. In this way, only the mean delay value of the interconnect is updated as a result of crosstalk effects, as shown in Figure 5.7.

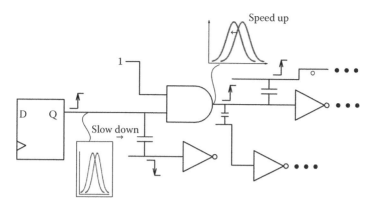

FIGURE 5.7
Cross talk impact on interconnect PDFs.

5.3 TDF Pattern Evaluation and Selection

In the proposed procedure, each TDF pattern is evaluated based on the paths it sensitizes. If a TDF pattern sensitizes a large number of long paths in the design, it would be considered an effective pattern. An in-house tool was developed to list all the sensitized paths for each pattern. Before the pattern evaluation and selection procedure, each sensitized path is calculated and evaluated in the presence of process variations and cross talk.

5.3.1 Path PDF Analysis

As mentioned, Monte Carlo simulation is used to obtain the PDFs for each path segment. The gate and interconnect delays are updated after considering process variations and crosstalk effects. With this information, the PDFs of the sensitized paths can be calculated. Each path is evaluated by comparing its length to the clock period. During pattern analysis, the clock period T is first obtained, and a long-path threshold LP_{thr} based on the clock period is specified to define the weight of the sensitized path W_{path}. The path weight is the probability that the path is longer than the predefined long-path threshold. Figure 5.8 shows an example of path weight definition. Using this method, each sensitized path is calculated and evaluated.

After calculating the weight of each sensitized path for *pattern$_i$*, the weight of *pattern$_i$* $W_{patterni}$ is calculated using Equation 6.6:

$$W_{patterni} = \sum_{j=1}^{M_i} W_{path\_ij} \qquad (6.6)$$

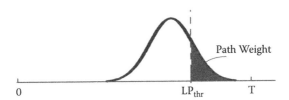

FIGURE 5.8
Path PDF and path weight definition.

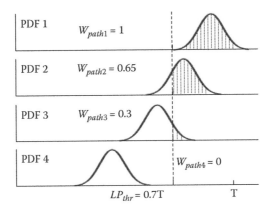

FIGURE 5.9
An example of path and pattern evaluation.

where M_i is the total number of sensitized paths by *pattern$_i$*, W_{path_ij} is the *j*th sensitized path by *pattern$_i$*.

Figure 5.9 shows an example of path weight and pattern weight calculation. In this example, it was assumed that *pattern$_i$* sensitized four different paths. The PDF of each path is shown in Figure 5.9 as *PDF1*, *PDF2*, *PDF3*, and *PDF4*, assuming that LP_{thr} was 0.7T. The weight of these four paths can be calculated as $W_{path1} = 1$, $W_{path2} = 0.65$, $W_{path3} = 0.3$, and $W_{path4} = 0$. Then, the weight of this pattern can be calculated according to Equation 6.6: $W_{patterni} = 1 + 0.65 + 0.3 + 0 = 1.95$.

5.3.2 Pattern Selection

From this analysis, it can be concluded that if a pattern has a large weight, it is more effective in detecting SDDs. Therefore, the patterns with largest weights will be selected for screening SDDs. However, some of the paths may be sensitized by multiple patterns. In the proposed procedure, if a path has already been detected by the selected patterns, it will not be considered during evaluation of the remaining patterns.

In this pattern selection procedure, the pattern with the largest weight in the pattern repository will be the first to be selected. After selecting

```
01: S ← pattern list of all the patterns, LP_thr ← long path threshold
02: S = {P_1, P_2, ·, P_N}
03: W(P_i) ← weight of pattern i, P_i based on LP_thr
04: P_s = NULL
05:    for (i = 1, · · · , N)
06:    {
07:       for (j = i, · · · , N)
08:       {
09:          update W(P_j) by removing paths detected by P_s
10:       }
11:       P_max = MAX{P_j, · · · , P_N}
12:       if (W(P_i) < W(P_max))
13:       {
14:          exchange P_i and P_max in pattern list S
15:       }
16:       P_s = P_i
17:    }
18: Return sorted pattern list for pattern selection.
```

FIGURE 5.10
Pattern-sorting algorithm.

the pattern, all the remaining patterns are reevaluated by excluding paths detected by the selected pattern. Then, the pattern with the largest weight in the remaining pattern set is selected. This procedure is repeated until the stopping criterion is met (i.e., the pattern weight is smaller than a specific threshold). This pattern selection procedure will ensure that the best patterns can be selected from the original pattern repository. It can also ensure that there is as little overlap as possible between patterns in terms of sensitized paths. Therefore, it can reduce the pattern count. The pattern-sorting algorithm is shown in Figure 5.10. This algorithm will return a pattern list with decreasing order according to the test efficiency of patterns with which the best patterns in the original pattern repository can be selected.

Assume that N patterns are used by the algorithm. Also assume that a maximum of M paths are sensitized by a pattern and a maximum of K segments exist on a sensitized path in the target circuit. The worst-case time complexity of the pattern sorting algorithm is $O(N^2MK)$ where $N \gg M$ and $N \gg K$ for large designs.

After pattern selection, top-off ATPG is run to meet the fault coverage requirement for the TDF fault model. As a result, the final pattern set of the proposed procedure is the selected pattern set plus the top-off ATPG pattern set. Figure 5.11 presents the entire flow of pattern evaluation and selection procedure. In this flow, commercial electronic design automation (EDA) tools were used for circuit synthesis, physical design, parasitic parameter extraction, as well as pattern generation. The programs for performing process variations and crosstalk calculation and pattern evaluation and selection were implemented using C/C++.

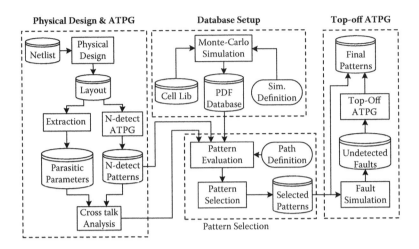

FIGURE 5.11
Flow diagram of pattern generation, evaluation, and selection.

TABLE 5.1

Details of Experimental Benchmarks

Benchmark	Number of Gates	Number of FFs*	Number of Total Cells
tv80	8,648	773	9,421
ac97_ctrl	25,488	3,034	28,522
mem_ctrl	13,388	1,705	15,093
systemcaes	12,529	1,584	14,113
wb_dma	8,474	1,398	9,872
s13207	1,355	625	1,980
s9234	511	145	656

* flip flops

5.4 Experimental Results and Analysis

The proposed procedure was implemented on several academic benchmarks to validate its efficiency. The details of these benchmarks are shown in Table 5.1. The number of logic gates is shown in column 2, and the number of flip-flops is shown in column 3. The total number of standard cells is presented in column 4. Note that the data in Table 5.1 were obtained from synthesized circuits, and they may be slightly different after placement and routing since the tool may add some buffers for routing optimization.

5.4.1 Pattern Selection Efficiency Analysis

The pattern selection efficiency analysis is performed by calculating the total number of sensitized unique long paths for the selected patterns. In

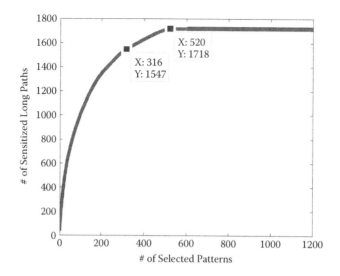

FIGURE 5.12
Comparison between the sensitized long paths and the number of selected patterns for tv80.

these experiments, the long-path threshold LP_{thr} was set at $0.7T$ (T is the clock period). A path was considered long if its weight was larger than 0.5, that is, its mean value was larger than the long-path threshold LP_{thr}.

Experimental results demonstrated that the pattern selection procedure was efficient in terms of selecting high-quality patterns for SDD detection. Figure 5.12 presents the relation between the number of selected patterns and sensitized unique long paths by the selected pattern set on IWLS benchmark tv80 (an 8-bit microprocessor core). For this benchmark, the *10-detect* pattern set with pattern count 11,072 was used as the original pattern repository. The number of sensitized long paths of this pattern repository is 1,718, with long-path threshold $LP_{thr} = 0.7T$. From the experimental results in Figure 5.12, it can be seen that only 520 (4.7% of the total pattern count 11,072) patterns were needed to detect all the long paths sensitized by the *10-detect* pattern set, or 316 (2.8% of the total pattern count 11,072) patterns were needed to detect 1,547 (or 90.0% of) long paths sensitized by the *10-detect* pattern set. For other benchmarks, the experimental results were similar to the tv80 benchmark results.

5.4.2 Pattern Set Analysis

In the pattern selection procedure, a pattern weight threshold is needed as the termination criterion. If the pattern weight threshold is 0, all the patterns will be selected. On the other hand, if the pattern weight threshold is even larger than the maximum weight of the patterns, no pattern can be selected. In the following experiments, the pattern weight threshold was $W_{pattern} = 1$, that is, only the patterns with weight larger than 1 could be

TABLE 5.2

Number of Sensitized Long Paths for Different Pattern Sets

Benchmark	1-detect	10-detect	t.aware	selected	top-off	sel+top-off
tv80	976	1,718	1,578	1,623	613	2,236
ac97_ctrl	343	400	360	373	92	465
mem_ctrl	237	2,739	1,168	2,707	186	2,893
systemcaes	772	2,110	1,763	1,940	28	1,968
wb_dma	1,036	1,890	1,209	1,881	98	1,979
s13207	143	155	136	152	25	177
s9234	172	309	203	305	3	308

TABLE 5.3

Number of Detected SDDs for Different Pattern Sets

Benchmark	1-detect	10-detect	t.aware	selected	top-off	sel+top-off
tv80	17,420	32,856	34,199	31,080	10,171	41,251
ac97_ctrl	3,825	4,704	4,183	4,376	1,007	5,383
mem_ctrl	4,614	53,650	22,156	53,024	3,416	56,440
systemcaes	13,366	38,380	32,988	35,334	440	35,774
wb_dma	13,842	26,701	20,131	26,594	1,299	27,893
s13207	2,167	2,629	1,881	2,682	306	2,988
s9234	2,128	3,724	2,442	3,679	44	3,723

selected. Changing this threshold can have an impact on the total number of selected patterns. Also in these experiments, the *10-detect* pattern set was used as the original pattern repository. The selected patterns as well as the following top-off patterns were compared with *1-detect*, *10-detect*, and timing-aware pattern sets in terms of long-path sensitization, SDD detection, and pattern count.

Tables 5.2 and 5.3 show the results for the number of sensitized unique long paths and detected unique SDDs. In general, *n-detect* and timing-aware pattern sets are expected to perform better in sensitizing unique long paths and detecting unique SDDs compared to *1-detect* timing-unaware ATPG. This is indicated by the results shown in columns 2, 3, and 4 in both tables. Column 5 presents the number of sensitized unique long paths of the selected pattern set. Column 6 presents the number of unique long paths sensitized by the top-off ATPG pattern set and not sensitized by the selected pattern set. Top-off patterns were generated using *1-detect* timing-unaware ATPG. Column 7 presents the total number of sensitized unique long paths for the final pattern set, that is, the selected patterns plus the top-off ATPG patterns.

From the results in Table 5.2, it can be seen that timing-aware ATPG and *10-detect* ATPG pattern sets always detected a significantly higher number of long paths than the *1-detect* pattern set except for the s13207 benchmark. On the other hand, timing-aware ATPG was not as effective as *10-detect* ATPG

TABLE 5.4

Pattern Count for Different Pattern Sets

Benchmark	1-detect	10-detect	t.aware	selected	top-off	sel+top-off
tv80	1,435	11,072	17,107	393	924	1,317
ac97_ctrl	1,032	6,393	4,087	126	834	960
mem_ctrl	1,595	13,142	6,577	352	1,032	1,384
systemcaes	591	3,686	5,590	800	30	830
wb_dma	483	3,600	4,460	131	354	485
s13207	810	6,712	1,108	36	775	811
s9234	343	2,763	428	64	271	335

in long-path sensitization for these circuits. However, the pattern set generated with the proposed procedure was more efficient than the *10-detect* ATPG pattern set in terms of long-path sensitization, except for the s9234 and systemcaes benchmarks.

Table 5.3 presents the number of SDDs for different pattern sets. Since SDDs are TDFs on the long paths, if a pattern detects many long paths, then it can also detect many SDDs. Table 5.4 presents the number of patterns for *1-detect*, *10-detect*, timing-aware ATPG pattern sets, and the pattern set from the proposed flow. These patterns were used for obtaining the number of sensitized long paths and detected SDDs as shown in Tables 5.2 and 5.3. From these results, it can be seen that timing-aware ATPG resulted in a large pattern count compared to *1-detect* set for large IWLS benchmarks. For some cases (e.g., tv80 and wb_dma benchmarks), its pattern count was even larger than the *10-detect* pattern set. For all cases, the proposed pattern set would result in a significantly smaller number of patterns compared to *10-detect* and timing-aware pattern sets. In short, the pattern set generated with the proposed procedure could detect a large number of long paths with a pattern count close to the *1-detect* pattern set.

5.4.3 Long-Path Threshold Analysis

The long-path threshold LP_{thr} is an important parameter for this procedure. It determinates how many patterns are going to be selected and further how many top-off patterns will be generated. If the long-path threshold changes, the path weight calculation threshold will change accordingly. Although it will not have an impact on the effectiveness of the pattern selection procedure, it may have an impact on the number of selected patterns and number of detected long paths and SDDs. If the long-path threshold increases, the number of selected patterns decreases and the number of top-off ATPG patterns increases to meet the fault coverage requirement. On the other hand, if the long-path threshold is reduced, the number of selected patterns increases and top-off ATPG pattern count decreases. The number of sensitized long paths and SDDs will also change accordingly. The previous

TABLE 5.5

Long-Path Threshold Impact on Pattern Selection for tv80

LP_{thr}	Number of Selected Patterns	Number of Top-off Patterns	Total Number of Patterns	Number of LPs	Number of SDDs
0.7T	393	924	1,317	2,236	82,502
0.8T	60	1,279	1,339	1,765	63,428

results are only for a fixed long-path threshold LP_{thr} (0.7T). Table 5.5 presents the results for two different long-path thresholds (0.7T to 0.8T) for the tv80 benchmark. When the long-path threshold increases, the weight of each pattern decreases, and the number of selected pattern decreases as well.

5.4.4 CPU Run Time Analysis

Table 5.6 presents the CPU run time of implementing the proposed method method on *n-detect* pattern sets (n = 1, 3, 5, 8, and 10) for the tv80 benchmark. It can be seen that as n increased, the pattern count increased. The CPU run time of the pattern evaluation and selection procedure also increased with the pattern count when considering (1) only process variations (next-to-last row in Table 5.6) and (2) process variations and cross talk (last row in Table 5.6). Furthermore, CPU run time increased significantly when considering crosstalk effects. This is because, during crosstalk calculation, for each net on the path, the procedure extracts (1) all its neighboring nets with coupling capacitances from the layout database, (2) the arrival time, and (3) transition directions on the neighboring nets (if any) for each pattern. This makes the pattern selection procedure much more complex.

Note that the timing-aware TDF ATPG on the tv80 benchmark took about 1 h 2 min, and the CPU time of *n-detect* timing-unaware TDF ATPG (n = 1, 3, 5, 8, and 10) on this benchmark was less than 2 min. As seen from the table, the pattern evaluation and selection procedure consumed considerably less CPU time when only process variations were considered. The top-off ATPG was quite fast and consumed negligible CPU time. When taking cross talk into consideration, the CPU time for evaluating the *5-detect* pattern set was close to that of timing-aware ATPG. Even though the proposed method was quite fast, comparing it with timing-aware ATPG does not seem fair since the

TABLE 5.6

CPU Run Time of Different Pattern Sets for tv80

n-detect	1-detect	3-detect	5-detect	8-detect	10-detect
Number of patterns	1,435	3,589	5,775	8,931	11,072
CPU (process variations)	48 s	2 min 17 s	4 min 2 s	6 min 32 s	8 min 3 s
CPU (process variations plus cross talk)	18 min 57 s	48 min 28 s	1 h 19 min 2 s	2 h 1 min 59 s	2 h 30 min 3 s

proposed method takes extra features into account during pattern generation. Furthermore, it is a comparison between experimental nonoptimized codes and the highly optimized commercial ATPG tool. With better data structures and programming, the CPU run time of this method can be further reduced.

5.5 Conclusion

In this chapter, the necessity of SDD detection was analyzed for higher test quality and low test escape, and a novel pattern evaluation and selection procedure for screening SDDs was presented. In the proposed procedure, each pattern is evaluated by its sensitized long paths. Process variations and cross talk are taken into account when evaluating the sensitized long paths. Note that the quality of the selected patterns depends on the original pattern repository; an *n-detect* pattern set was used as the pattern repository. This procedure was able to efficiently identify high-quality patterns in the pattern repository that sensitize a large number of long paths. The method was evaluated using several International Symposium on Circuits (ISCAS) and IWLS benchmarks, and the results demonstrated its effectiveness for reducing the pattern count and significantly increasing the number of sensitized long paths.

5.6 Acknowledgment

We would like to thank Prof. Krihsnendu Chakrabarty of Duke University for his contribution to this chapter.

References

[Ahmed 2006a] N. Ahmed, M. Tehranipoor, and V. Jayaram, A novel framework for faster-than-at-speed delay test considering IR-drop effects, in *Proceedings International Conference on Computer-Aided Design*, 2006, pp. 198–203.
[Ahmed 2006b] N. Ahmed, M. Tehranipoor, and V. Jayram, Timing-based delay test for screening small delay defects, in *Proceedings IEEE Design Automation Conference*, 2006.
[Amyeen 2004] M.E. Amyeen, S. Venkataraman, A. Ojha, and S. Lee, Evaluation of the quality of n-detect scan ATPG patterns on a processor, in *Proceedings IEEE International Test Conference [ITC'04]*, 2004, pp. 669–678.

[Chen 1997] W. Chen, S. Gupta, and M. Breuer, Analytic models for crosstalk delay and pulse analysis under non-ideal inputs, in *Proceedings IEEE International Test Conference*, 1997, pp. 808–818.

[Goel 2009] S.K. Goel, N. Devta-Prasanna, and R. Turakhia, Effective and efficient test pattern generation for small delay defects, in *Proceedings IEEE VLSI Test Symposium*, 2009.

[Gupta 2004] P. Gupta and M. Hsiao, ALAPTF: A new transition fault model and the ATPG algorithm, in *Proceedings IEEE International Test Conference*, 2004, pp. 1053–1060.

[ITRS 2008] International Technology Roadmap for Semiconductors home page, 2008, http://www.itrs.net/Links/2008ITRS/Home2008.htm

[Koralov 2007] L.B. Koralov and Y.G. Sinai, *Theory of Probability and Random Processes*, 2nd edition, Springer, New York, 2007.

[Lee 2006] H. Lee, S. Natarajan, S. Patil, and I. Pomeranz, Selecting high-quality delay tests for manufacturing test and debug, in *Proceedings IEEE International Symposium on Defect and Fault-Tolerance in VLSI Systems*, 2006.

[Lion 2002] J. Lion, A. Krstic, L. Wang, and K. Cheng, False-path-aware statistical timing analysis and efficient path selection for delay testing and timing validation, in *Proceedings IEEE Design Automation Conference*, 2002, pp. 566–569.

[Majhi 2000] A.K. Majhi, V.D. Agrawal, J. Jacob, and L.M. Patnaik, Line coverage of path delay faults, *IEEE Transactions on Very Large Scale Integration [VLSI] Systems* 8(5), pp. 610–614, 2000.

[Mattiuzzo 2009] R. Mattiuzzo, D. Appello, and C. Allsup, Small delay defect testing, *Test and Measurement World*, Lexington, MA (online) 2009. Available: http://www.tmworld.com/article/CA6660051.html

[Mentor 2006] *Understanding How to Run Timing-Aware ATPG*, Application Note, 2006.

[Qiu 2004] W. Qiu, J. Wang, D. Walker, D. Reddy, X. Lu, Z. Li, W. Shi, and H. Balachandran, *K* longest paths per gate [KLPG] test generation for scan-based sequential circuits, in *Proceedings IEEE International Test Conference*, 2004, pp. 223–231.

[Rabaey 2003] J.M. Rabaey, A. Chandrakasan, and B. Nikolic, *Digital Integrated Circuits, a Design Perspective*, 2nd edition, Prentice Hall, Upper Saddle River, NJ, 2003.

[Recktenwald 2000] G.W. Recktenwald, *Numerical Methods with MATLAB: Implementations and Applications*, Prentice-Hall, Upper Saddle River, NJ, 2000.

[Savir 1992] J. Savir, Skewed-load transition test: Part I, calculus, in *Proceedings IEEE International Test Conference*, 1992, pp. 705–713.

[Savir 1994b] J. Savir and S. Patil, On broad-side delay test, in *Proceedings IEEE VLSI Test Symposium*, 1994, pp. 284–290.

[Synopsys 2007a] Synopsys Inc., *SOLD Y-2007*, Volumes 1–3, Synopsys Inc., Mountain View, CA, October 2007.

[Yilmaz 2008b] M. Yilmaz, K. Chakrabarty, and M. Tehranipoor, Interconnect-aware and layout-oriented test-pattern selection for small-delay defects, in *Proceedings International Test Conference [ITC'08]*, 2008.

[Yilmaz 2008c] M. Yilmaz, K. Chakrabarty, and M. Tehranipoor, Test-pattern grading and pattern selection for small-delay defects, in *Proceedings IEEE VLSI Test Symposium*, 2008, pp. 233–239.

Section III

Alternative Methods

6

Output Deviations-Based SDD Testing

Mahmut Yilmaz

CONTENTS

6.1 Introduction

Very deep-submicron (VDSM) process technologies are leading to increasing densities and higher clock frequencies for integrated circuits (ICs). However, VDSM technologies are susceptible to process variations, crosstalk noise, power supply noise, and defects such as resistive shorts and opens, which induce small-delay variations in the circuit components. The literature refers to such delay variations as small-delay defects (SDDs) [Ahmed 2006b].

Although the delay introduced by each SDD is small, the overall impact might be significant if the target path is critical, has low slack, or includes many SDDs. The overall delay of the path might become longer than the clock period, causing circuit failure or temporarily incorrect results. As a result, the detection of SDDs requires fault excitation through least-slack paths. The longest paths in the circuit, except false paths and multicycle paths, are referred to as the least-slack paths.

6.2 The Need for Alternative Methods

The transition delay fault (TDF) [Waicukauski 1987] model attempts to propagate the lumped-delay defect of a gate by logical transitions to the observation points or state elements. The TDF model is not effective for SDDs because test generation using TDFs leads to the excitation of short paths [Ahmed 2006b; Forzan 2007]. Park et al. first proposed a statistical method for measuring the effectiveness of delay fault automatic test pattern generation (ATPG) [Park 1989]. The proposed technique is especially relevant today, and it can handle process variations on sensitized paths; however, this work is limited in the sense that it only provides a metric for delay test coverage, and it does not aim to generate or select effective patterns.

Due to the growing interest in SDDs, vendors recently introduced the first commercial timing-aware ATPG tools (e.g., new versions of Mentor Graphics FastScan, Cadence Encounter Test, and Synopsys TetraMax). These tools attempt to make ATPG patterns more effective for SDDs by exercising longer paths or applying patterns at higher-than-rated clock frequencies. However, only a limited amount of timing information is supplied to these tools, either via standard delay format (SDF) files (for FastScan and Encounter Test) or through a static timing analysis (STA) tool (for TetraMax). As a result, users cannot easily extend these tools to account for process variations, cross talk, power supply noise, or similar SDD-inducing effects on path delays. These tools simply rely on the assumption that the longest paths (determined using STA or SDF data) in a design are more prone to failure due to SDDs. Moreover, the test generation time increases considerably when running these tools in timing-aware mode.

Figure 6.1 shows a comparison of the run times of two ATPG tools from the same electronic design automation (EDA) company: (1) timing-unaware ATPG (i.e., a traditional TDF pattern generator) and (2) timing-aware ATPG. Figure 6.1 shows results for some representative AMD microprocessor functional blocks. Figure 6.1 demonstrates as much as 22 times longer CPU (central processing unit) run time and 15 times greater pattern count. This research normalized numbers to take the run time and pattern count of timing-unaware ATPG as one unit. Given that TDF pattern generation could take a couple of days or more for industrial designs, the run time of timing-aware ATPG is not feasible.

A path delay fault (PDF) ATPG is an alternative for generating delay test patterns for timing-sensitive paths. Since the number of paths in a large

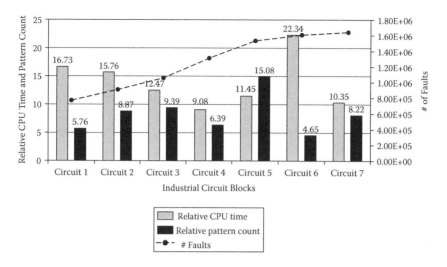

FIGURE 6.1
CPU run time and pattern count of timing-aware ATPG relative to traditional TDF ATPG for subblocks of an industrial microprocessor circuit.

design is prohibitively large, it is not practical to target all paths in the design. Instead, the PDF ATPG first identifies a set of candidate paths for test generation. A typical flow for PDF pattern generation consists of the following two steps [Padmanaban 2004]:

- Generate a list of timing-sensitive paths using an STA tool. Usually, the tool reports only the most critical paths (e.g., paths with less than 5% slack margin).

- Next, feed the reported set of paths into an ATPG tool to generate patterns targeting these paths.

There are, however, several drawbacks to this flow. First, the predicted timing data rarely match the real delays observed on silicon [Cheng 2000] due to process variations. As a result, if we target only a small list of critical paths, we are likely to miss real critical delay paths due to timing variations and secondary effects such as cross talk. As reported [Zolotov 2010], all semiconductor companies recognize this problem. One solution is to report a larger group of paths by increasing the slack margin. For instance, one might report paths with 20% slack margin to account for delay variations. This approach, however, introduces another drawback of PDF ATPG flow: The number of paths increases exponentially as the slack margin is increased. Hence, increased slack margins are also impractical for large designs.

We experimented with production circuits to quantify this problem. We observed that, for various industrial designs, the number of paths reported for a 5% slack margin was more than 100,000. The ATPG tool generated twice the number of patterns compared with 1-detect TDF patterns, and it could test less than half of all PDFs. (All other paths remained untested.) Increasing the slack margin to 10% pushed the number of paths to millions and effectively made it infeasible to run the PDF ATPG. Results for slack margin limits as high as 20% are practically impossible to report due to disk space limitations. This highlights the problems of using PDF ATPG for large industrial circuits fabricated using nanoscale complementary metal oxide semiconductor (CMOS) processes that are subject to process variations and provide some quantitative data points.

Statistical static timing analysis (SSTA) can generate variability-aware delay data. Although a complete SSTA flow takes considerable computation time [Forzan 2007; Nitta 2007], simplified SSTA-based approaches could be used for pattern selection, as shown by some authors [Chao 2004; Lee 2005]. Chao and coworkers [Chao 2004] proposed an SSTA-based test pattern quality metric to detect SDDs. The computation of the metric required multiple dynamic timing analysis runs for each test pattern using randomly sampled delay data from Gaussian pin-to-pin delay distributions. The proposed metric was also used for pattern selection. Other authors [Lee 2005] focused on timing hazards and proposed a timing hazard-aware SSTA-based pattern selection technique.

The complexity of today's ICs and shrinking process technologies also leads to prohibitively high test data volumes. For example, the test data volume for TDFs is two to five times higher than stuck-at faults [Keller 2004], and it has also been demonstrated that test patterns for such sequence- and timing-dependent faults are more important for newer technologies [Ferhani 2006]. The 2007 International Technology Roadmap for Semiconductors (ITRS) predicted that the test data volume for ICs will be as much as 38 times larger and the test application time will be about 17 times longer in 2015 than it was in 2007. Therefore, efficient generation, pattern grading, and selection methods are required to reduce the total pattern count while effectively targeting SDDs.

This chapter presents a statistical approach based on the concept of output deviations to select the most effective test patterns for SDD detection [Yilmaz 2008c, 2010]. The literature indicates that, for academic benchmark circuits, there is high correlation between the output deviations measure for test patterns and the sensitization of long paths under process variation. These benchmark circuits report significant reductions in CPU run time and pattern count and higher test quality compared with commercial timing-aware ATPG tools. We also examine the applicability or effectiveness of this approach for realistic industrial circuits. We present the adaptation of the output deviation metric to industrial circuits. This work enhanced the framework of output deviations to make it applicable for such circuits.

Experimental results showed the proposed method could select the highest-quality patterns effectively from test sets too large for use in their entirety in production test environments with tight pattern count limits. Unlike common industry ATPG practices, it also considers delay faults caused by process variations. The proposed approach incurs a negligible runtime penalty. For two important quality metrics—coverage of long paths and long-path coverage ramp up—this approach outperforms a commercial timing-aware ATPG tool for the same pattern count.

6.3 Probabilistic Delay Fault Model and Output Deviations for SDDs

This section reviews the concept of output deviations, then describes how we extended the deviation-based pattern selection method of earlier work to industrial circuits.

6.3.1 Method of Output Deviations

Yilmaz and coworkers [Yilmaz 2008c] introduced the concepts of gate delay defect probabilities (DDPs) and signal transition probabilities (STPs). These probabilities introduce the notion of confidence levels for pattern pairs.

In this section, we first review the concept of DDPs (Section 6.3.1.1) and STPs (Section 6.3.1.2). These probabilities extend the notion of confidence levels, defined by Wang et al. [Wang 2008a] for a single pattern, to pattern pairs. Next, we show how to use these probability values to propagate the effects of a test pattern to the test observation points (scan flip-flops/primary outputs) (Section 6.3.1.3). We describe the algorithm used for signal probability propagation (Section 6.3.1.4). Finally, we describe how to rank and select test patterns from a large repository (Section 6.3.1.5).

6.3.1.1 Gate Delay Defect Probabilities

This approach assigns DDPs to each gate in a design, providing DDPs for a gate in the form of a matrix called the delay defect probability matrix (DDPM). Table 6.1a shows the DDPM for a two-input OR gate. The rows in the matrix correspond to each input port of the gate, and the columns correspond to the initial input state during a transition.

Assume the inputs are shown in the order IN0, IN1. If there is an input transition from 10 to 00, the corresponding DDPM column is 10. Since IN0 causes the transition, the corresponding DDPM row is IN0. As a result, the DDP value corresponding to this event is 0.5, showing the probability that the corresponding output transition is delayed beyond a threshold.

TABLE 6.1

Example DDPM for a Two-Input OR Gate: (a) Traditional Style and (b) New Style

		(a)				(b)				
		Initial Input State				$L \rightarrow H$		$H \rightarrow L$		
	Probability	00	01	10	11		μ	σ^2	μ	σ^2
Inputs	IN0	0.21	0	0.5	0.11	IN0 $\rightarrow$ Z	100 ps	112	110 ps	97
	IN1	0.12	0.20	0		IN1 $\rightarrow$ Z	110 ps	123	120 ps	117

For initial state 11, both inputs should switch simultaneously to have an output transition. This approach merges corresponding DDPM entries due to this requirement. We chose the entries in Table 6.1 arbitrarily for the sake of illustration. The real DDPM entries are much smaller than those shown in this example.

For an N-input gate, the DDPM consists of $N * 2^N$ entries, each holding one probability value. If the gate has more than one output, each output of the gate has a different DDPM. Note that the DDP is 0 if the corresponding event cannot provide an output transition. Consider DDPM(2,3) in Table 6.1. When the initial input state is 10, no change in IN1 might cause an output transition because the OR gate output is already at high state; even if IN1 switches to high (1), this will not cause an output transition.

We next discuss how to generate a DDPM. Each entry in a DDPM indicates the probability that the delay of a gate is more than a predetermined value (i.e., the critical delay value T_{CRT}). Given the probability density function of a delay distribution, the DDP is calculated as the probability that the delay is higher than T_{CRT}.

For instance, if we assume a Gaussian delay distribution for all gates (with mean μ) and set the critical delay value to $\mu + X$, each DDP entry can be calculated by replacing T_{CRT} with $\mu + X$ and using a Gaussian probability density function. Note that the delay for each input-to-output transition could have a different mean μ and standard deviation σ.

We can obtain the delay distribution in different ways: (1) using the delay information provided by an SSTA-generated SDF file; (2) using slow, nominal, and fast process corner transistor models; or (3) simulating process variations. In the third method, which we apply in this chapter to academic benchmarks, we first determined transistor parameters affecting the process variation and the limits of the process variation (3σ). Next, we ran Monte Carlo simulations for each library gate under different capacitive loading and input slew rate conditions. Once this process finds distributions for the library gates, depending on the layout, it can update the delay distributions for each individual gate. Once this process obtains distributions, it can set T_{CRT} appropriately to compute the DDPM entries. We can simulate the effects of cross talk separately, updating the delay distributions of individual gates/wires accordingly.

The generation of the DDPMs is not the main focus of this chapter. We consider DDPMs analogous to timing libraries. Our goal is not to develop the most effective techniques for constructing DDPMs; rather, we are using such statistical data to compute deviations and use them for pattern grading and pattern selection. In a standard industrial flow, specialized timing groups develop statistical timing data, so the generation of DDPMs is a preprocessing step and an input to the ATPG-focused test automation flow.

We have seen that small changes in the DDPM entries (e.g., less than 15%) have a negligible impact on the pattern selection results. We attribute this finding to the fact that any DDPM changes affect multiple paths in the circuits, thus amortizing their impact across the circuit and the test set. The absolute values of the output deviations are less important than the relative values for different test patterns.

6.3.1.2 Propagation of Signal Transition Probabilities

Since pattern pairs are required to detect TDFs, there might be a transition on each net of the circuit for every pattern pair. If we assume there are only two possible logic values for a net (i.e., LOW [L] and HIGH [H]), the possible signal transitions are $L \rightarrow L$, $L \rightarrow H$, $H \rightarrow L$, and $H \rightarrow H$. Each of these transitions has a corresponding probability, denoted by $P_{L \rightarrow L}$, $P_{L \rightarrow H}$, $P_{H \rightarrow L}$, and $P_{H \rightarrow H}$, respectively, in a vector form ($<...>$): $< P_{L \rightarrow L}, P_{L \rightarrow H}, P_{H \rightarrow L}, P_{H \rightarrow H} >$. We refer to this vector as the STP vector. Note that $L \rightarrow L$ or $H \rightarrow H$ implies that the net keeps its value (i.e., no transition occurs).

The nets directly connected to the test application points are called initialization nets (INs). These nets have one of the STPs, corresponding to the applied transition test pattern, equal to 1. All the other STPs for INs are set to 0. When signals propagate through several levels of gates, this approach can compute STPs using the DDPM of the gates. Note that interconnects can also have DDPMs to account for cross talk. In this chapter, due to the lack of layout information, we only focus on variations' impact on gate delay. The overall deviation-based framework is, however, general, and it could easily accommodate interconnect delay variations if layout information is available, as reported by Yilmaz et al. [Yilmaz 2008b].

Definition 1

Let P_E be the probability that a net has the expected signal transition. The deviation on that net is defined by $\Delta = 1 - P_E$. We apply the following rules during the propagation of STPs:

1. If there is no output signal transition (output keeps its logic value), then the deviation on the output net is 0.

2. If there are multiple inputs that might cause the expected signal transition at the output of a gate, only the input-to-output path that causes the highest deviation at the output net is considered. The other inputs are treated as if they have no effect on the deviation calculation (i.e., they are held at the noncontrolling value).
3. When multiple inputs must change at the same time to provide the expected output transition, we consider all required input-to-output paths of the gate. We discard only the unnecessary (redundant) paths.

A key premise of this chapter is that we can use output deviations to compare path lengths. As in the case of path delays, the net deviations also increase as the signal propagates through a sensitized path, a property that follows from the rules used to calculate STPs for a gate output. Next, we formally prove this claim.

Lemma 1

For any net, let the STP vector be given by $< P_{L \to L}, P_{L \to H}, P_{H \to L}, P_{H \to H} >$. Among these four probabilities (i.e., $< P_{L \to L}, P_{L \to H}, P_{H \to L}, P_{H \to H} >$), at least one is nonzero, and at most two can be nonzero.

PROOF

If there is no signal value change (the event L→L or H→H), the expected STP is 1, and all other probabilities are 0. If there is a signal value change, only the expected signal transition events and the delay fault case have nonzero probabilities associated with them. The delay fault case for an expected signal value change of $L \to H$ is $L \to L$ (the signal value does not change because of a delay fault). Similarly, the delay fault case for an expected signal value change of $H \to L$ is $H \to H$. ∎

Theorem 1

The deviation on a net always increases or stays constant on a sensitized path if the signal probability propagation rules are applied.

PROOF

Consider a gate with K inputs and one output. The signal transition on the output net depends on one of the following cases: From Lemma 1, we need to consider only two cases:

i. Only one of the input port signal transitions is enough to create the output signal transition.

ii. Multiple input port signal transitions are required to create the output signal transition.

Let $P_{OUT,j}$ be the probability that the gate output makes the expected signal transition for a given pair of patterns on input j, where $1 \leq j \leq K$. Let $\Delta_{OUT,j} = 1 - P_{OUT,j}$ be the deviation for the net corresponding to the gate output.

Case (*i*)

Consider a signal transition on input j. Let Q_j be the probability of occurrence of this transition. Let d_j be the entry in the gate's DDPM that corresponds to the given signal transition on j. The probability that the output makes a signal transition is

$$P_{OUT,j} = Q_j(1 - d_j)$$

We assume an error at a gate input is independent of the error introduced by the gate. Note $P_{OUT,j} \leq Q_j$ since $0 \leq d_j \leq 1$. Therefore, the probability of getting the expected signal transition decreases, and the deviation $\Delta_{OUT,j} = 1 - P_{OUT,j}$ increases (or does not change) as we pass through a gate on a sensitized path. The overall output deviation Δ_{OUT}^* on the output net is

$$\Delta_{OUT}^* = \max_{i \leq j \leq K}\{\Delta_{OUT,j}\} \qquad \blacksquare$$

Case (*ii*)

Suppose L input ports ($L > 1$), indexed 1, 2, ... , L, are required to make a transition for the gate output to change. Let $d_{max}^* = \max_{1 \leq j \leq L}\{d_j\}$. The output deviation for the gate in this case is

$$\Delta_{OUT}^* = \Pi\, P_{OUT,i} * (1 - d_{max}^*),\, 1 \leq i \leq L$$

Note that $\Delta_{OUT}^* \leq P_{OUT,i}$, $1 \leq i \leq L$ since $0 \leq d_{max}^* \leq 1$. Therefore, we conclude that the probability of getting the expected transition on a net either decreases or remains the same as we pass through a logic gate. In other words, the deviation is monotonically nondecreasing along a sensitized path. $\qquad \blacksquare$

Example

Figure 6.2 shows STPs and their propagation for a simple circuit. The dark boxes show test stimuli and the expected fault-free transitions on each net. The angled brackets (< ... >) show calculated STPs. Tables 6.1 and 6.2 give the DDPMs of the gates (OR, AND, XOR, and INV) used in

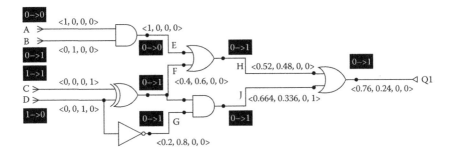

FIGURE 6.2
Example of the propagation of STPs through the gates of a logic circuit.

TABLE 6.2

Example DDPM for AND, XOR, INV

		Initial Input State			
AND		**00**	**01**	**10**	**11**
Inputs	IN0	0.2	0.3	0	0.2
	IN1		0	0.2	0.3
XOR		**00**	**01**	**10**	**11**
Inputs	IN0	0.3	0.4	0.1	0.2
	IN1	0.3	0.4	0.2	0.4
INV		**0**	**1**		
Input	IN0	0.2	0.2		

this circuit. The entries in both tables are arbitrary. We calculated the deviations in the following example based on the rules previously mentioned for the example circuit in Figure 6.2:

- **Net E:** There is no output change, which implies that E has the STP <1,0,0,0>.
- **Net F:** The output changes due to IN1 (net D) of XOR. There is a DDP of 0.4. It implies that, with a probability of 0.4, the output will stay at LOW value (i.e., the STP for net F is <0.4,0.6,0,0>).
- **Net G:** Output changes due to IN0 (net D) of INV (i.e., the STP for net G is <0.2,0.8,0,0>).
- **Net H:** Output changes due to IN1 (net F) of OR.
  - If IN1 stays at LOW, output does not change. Therefore, the STP for net H is 0.4 • <1,0,0,0>, where • denotes the dot product;
  - If IN1 goes to HIGH, output changes with a DDP of 0.2 (i.e., the STP for net H is 0.6 • <0.2,0.8,0,0>);
  - Combining all the preceding cases, the STP for net H is <0.52,0.48,0,0>.
- **Net J:** Output changes due to both IN0 (net F) and IN1 (net G) of AND (both required).

- If both stay at LOW, output does not change, which implies the STP for net J is 0.4 • 0.2 • <1,0,0,0>;
- If one stays at LOW, output does not change (i.e., the STP for net J is 0.4 • 0.8 • <1,0,0,0> + 0.6 • 0.2 • <1,0,0,0>);
- If both go to HIGH, the output changes with a DDP. Since both inputs change, we use the maximum DDP (i.e., the STP for net J is 0.6 • 0.8 • <0.3,0.7,0,0>);
- Combining all the preceding cases, the STP for net J is <0.664,0.336,0,0>.
- **Net Q1:** The output changes due to only one of the inputs of OR. We need to calculate the deviation for both cases and select the one that causes maximum deviation at the output (Q1).
  - For IN0 (net H) of OR:
    - If IN0 stays at LOW, the output does not change (i.e., the STP for net Q1 is 0.52 • <1,0,0,0>);
    - If IN0 goes to HIGH, the output changes with a DDP (i.e., the STP for net Q1 is 0.48 • <0.5,0.5,0,0>);
    - Combining all the preceding cases, the STP for net Q1 is <0.76,0.24,0,0>.
  - For IN1 (net J) of OR:
    - If IN1 stays at LOW, the output does not change (i.e., the STP for net Q1 is 0.664 • <1,0,0,0>);
    - If IN1 goes to HIGH, the output changes with a DDP (i.e., the STP for net Q1 is 0.336 • <0.2,0.8,0,0>);
    - Combining all the preceding cases, the STP for net Q1 is <0.7312,0.2688,0,0>.
  - Since IN0 provided the higher deviation, we finally conclude that the STP for net Q1 is <0.76,0.24,0,0>.

Hence, the deviation on Q1 is 0.76.

6.3.1.3 Implementation of Algorithm for Propagating Signal Transition Probabilities

We use a depth-first procedure to compute STPs for large circuits. A depth-first algorithm processes only the nets required to find the output deviation on a specific observation point. In this way, less gate pointer stacking is required relative to simulating the deviations starting from INs and tracing forward.

We first assign STPs to all INs. Then, we start from the observation points (outputs) and backtrace until we find a processed net (PN). A PN has all the STPs assigned. Figure 6.3 gives the pseudocode for the algorithm.

If the number of test patterns is N_s and the number of nets in the circuit is N_n, the worst-case time complexity of the algorithm is $O(N_s N_n)$. However, we could easily make the algorithm multithreaded because the calculation for each pattern is independent of other patterns (we assume full-scan designs in this chapter). In this case, if the number of threads is T, the complexity of the algorithm is reduced to $O((N_s N_n)/T)$.

Procedure: Propagate Probability (t_i)
1: reset all signal-transition probabilities
2: read pattern t_i
3: assign signal-transition probabilities to INs
4: reset stack
5: **for all** observation points PO_j. $j = 1, 2, ..$ **do**
6:      **if** PO_j is processed **then**
7:          go to next PO_j
8:      **end if**
9:      trace backward until a processed net is found
10:     add unprocessed gates on the traced path to the stack
11:     **for all** G =gate in stack **do**
12:         find signal-transition probabilities of the output net of G
13:         remove G from the stack
14:     **end for**
15:     find signal-transition probabilities of PO_j
16: **end for**

FIGURE 6.3
Signal transition probability propagation algorithm for calculating output deviations.

6.3.1.4 Pattern-Selection Method

In this section, we describe how to use output deviations to select high-quality patterns from an n-detect transition fault pattern set. The number of test patterns to select is a user input (e.g., S). The user can set the parameter S to the number of 1-detect timing-unaware patterns, the number of timing-aware patterns, or any other value that fits the user's test budget.

In our pattern selection method, we target topological coverage as well as long-path coverage. As a result, we attempt to select patterns that sensitize a wide range of distinct long paths. In this process, we also discard low-quality patterns to find a small set of high-quality patterns.

For each test observation point PO_j, we keep a list of N_p most effective patterns in EFF_j (Figure 6.4, lines 1–3). The patterns in EFF_j are the best unique-pattern candidates for exciting a long path through PO_j. During deviation computation, no pattern t_i is added to EFF_j if the output deviation at PO_j is smaller than a limit ratio D_{LIMIT} of the maximum instantaneous output deviation (Figure 6.4, line 10). We can use D_{LIMIT} to discard low-quality patterns. If the output deviation is larger than this limit, we first check whether we have added a pattern to EFF_j with the same deviation (Figure 6.4, line 11). It is unlikely that two different patterns will create the same output deviation on the same output PO_j while exciting different nonredundant paths. Since we want higher topological path coverage, we skip these cases (Figure 6.4, line 11). Although this assumption might not necessarily be true, we assume for the sake of completeness that it holds for most cases.

If we observed a unique deviation on PO_j, we first check whether EFF_j is full (whether it already includes N_p patterns; see Figure 6.4, line 12). Pattern t_i is added to EFF_j along with its deviation if EFF_j is not full or if t_i has a greater deviation than the minimum deviation stored in EFF_j (Figure 6.4, lines 12–17). We measure the effectiveness of a pattern by the number of occurrences of

```
Procedure: Compute Deviations (t_0, ..., t_{N_s}, N_p)
1:  for all observation point PO_j, j = 1, 2, ... do
2:      create list EFF_j[N_p]
3:  end for
4:  Max_Dev = 0;
5:  for all test pattern t_i, i = 1, 2, ..., N_s do
6:      Propagate Probability(t_i);
7:      for all observation point PO_j, j = 1, 2, ... do
8:          Dev = deviation of PO_j;
9:          if Dev > Max_Dev then Max_Dev = Dev;
10:         if Dev > D_{LIMIT}·Max_Dev then
11:             if EFF_j includes Dev then Next observation point;
12:             if EFF_j is not full then
13:                 add t_i and Dev to EFF_j;
14:             else if Dev > min(EFF_j) then
15:                 remove min(EFF_j);
16:                 add t_i and Dev to EFF_j;
17:             end if
18:         end if
19:     end for
20: end for
```

FIGURE 6.4
Deviation computation algorithm for pattern selection.

this pattern in EFF_j for all values of j. For instance, if—at the end of deviation computation—pattern A was included in the EFF list for 10 observation points and pattern B was listed in the EFF list for 8 observation points, we conclude that pattern A is more effective than pattern B.

Once the deviation computation is completed, this approach generates the list of pattern effectiveness and carries out the final pattern filtering and selection (Figure 6.5). First, pattern effectiveness is generated (Figure 6.5, lines 1–9). Since Max_Dev updates on the fly, we could miss some low-quality patterns. As a result, we need to filter by Max_Dev (D_{LIMIT}) again to discard low-quality patterns from the final pattern list (Figure 6.5, line 5). Setting D_{LIMIT} to a high value might result in discarding most of the patterns, leaving only the best few patterns. Depending on D_{LIMIT}, the number of remaining patterns might be less than S.

```
Procedure: Select Patterns D_S(t_0, ..., t_{N_s}, S, D_{LIMIT})
1:  list D[N_s];
2:  init D to all 0s;
3:  for all test pattern t_i, i = 1, 2, ..., N_s do
4:      for all observation point PO_j, j = 1, 2, ... do
5:          if EFF_j includes t_i and deviation of t_i > D_{LIMIT}·
            Max_Dev then
6:              increment D[i];
7:          end if
8:      end for
9:  end for
10: sort D by values;
11: D_S = select top S patterns;
12: return D_S;
```

FIGURE 6.5
Pattern selection algorithm.

In the next stage, this approach re-sorts the patterns by their effectiveness (Figure 6.5, line 10). Finally, until the selected pattern number reaches S or this approach selects all patterns, the top patterns are selected (Figure 6.5, line 11). The computational complexity of the selection algorithm is $O(N_s p)$, where N_s is the number of test patterns, and p is the number of observation points. This procedure is very fast since it only includes two nested for loops and a simple list item existence check.

6.3.2 Practical Aspects and Adaptation to Industrial Circuits

We need to revisit the input data required to compute output deviations to ensure provision of appropriate information, and we can use the proposed approach with the data available in an industrial project.

The two most significant inputs required by the previously proposed output deviations method are the gate and interconnect delay variations and T_{CRT} for gates and interconnects. Typically, a timing group computes delay variations for library gates based on design-for-manufacturability (DfM) rules. The available data are the input-to-output delay values for worst, nominal, and best conditions. Delay variations for interconnects are computed based on DfM rules defining the range of resistance and capacitance variations for different metal layers and vias.

The main challenge in practice is finding a specific T_{CRT} for gates and interconnects. Defining a T_{CRT} for individual gates and interconnects is not feasible for industrial circuits because allowed delay ranges are defined at the circuit or subcircuit levels. Due to this limitation, it was not possible to generate DDPM tables for gates and interconnects. As a result, we redefined the manner in which this approach computed output deviations.

We first assumed independent Gaussian delay distributions for each path segment (i.e., gates and interconnects), where nominal delay was used as the mean, and the worst-case delay was used as 3σ. Instead of using specific probability values in DDPMs, we used mean delay and variances for each gate instance and interconnect. An example of the new DDPM (with entries chosen arbitrarily) for a two-input OR gate is shown in Table 6.1b. The rows in the matrix correspond to each input-to-output timing arc, and the columns correspond to the mean μ and variance σ^2 of the corresponding $L \to H$ (rising) or $H \to L$ (falling) output transition.

Similarly, instead of propagating the STPs, we propagated the mean delay and variance on each path using the central limit theorem (CLT) [Trivedi, 2001], similar to the method proposed by Park et al. [Park 1989].

Since we assumed independent Gaussian distributions, we can use the following equations to calculate the mean value and standard deviation of the path probability density function [Trivedi 2001].

$$\mu_c = \sum_{i=1}^{N} \mu_i$$

$$\sigma_c = \sqrt{\sum_{i=1}^{N} \sigma_i^2}$$

where μ_i and σ_i are the probability density function mean value and standard deviation of segment i, respectively; μ_c and σ_c are the path probability density function mean value and standard deviation, respectively; and N is the number of path segments.

Even if we do not assume Gaussian distributions for each delay segment, as long as segment delays are independent distributions with finite variances, the CLT ensures that the path delay—which is the sum of segment delays—converges to a Gaussian distribution [Trivedi 2001].

We defined T_{CRT} for the circuit as a fraction of the functional clock period T_{func} of the circuit. In our experiments, we used the values $0.7 * T_{func}$, $0.8 * T_{func}$, $0.9 * T_{func}$, and T_{func}. For each case, the output deviation is the probability that the calculated delay on an observation point (scan flip-flops or primary outputs) is larger than T_{CRT}.

We also adapted the pattern selection method by introducing a degree of path enumeration to the pattern selection procedure. We implemented this change to ensure selection of all patterns exciting the delay-sensitive paths. We developed an in-house tool to list all the sensitized paths for a TDF pattern, in addition to each segment of the sensitized paths. This tool enumerates all paths sensitized by a given test pattern. The steps in this flow are as follows:

- Use commercial tools for ATPG and fault simulation.

- For each pattern, the ATPG tool reports the detected TDFs. This report includes the net name as well as the type of signal transition (i.e., falling or rising).

- Our in-house tool finds the active nets (i.e., nets that have a signal transition on them) in the circuit under test. This step has $O(\log N)$ worst-case time complexity.

- Starting from scan flip-flops and primary inputs, trace forward each net with a detected fault. Note that, if a fault is detected, it means this net reaches a scan flip-flop through other nets with detected faults. If the sensitized path has no branching on it, the complexity of this step is $O(N)$. However, if there are K branches on the sensitized net, and if all these branches create a different sensitized path, the complexity of this step is $O(N^K)$.

- Note that, although unlikely, if a test pattern could test all nets at the same time, the run time of the sensitized path search procedure would be very high. However, our simulations on academic benchmarks and industrial circuits showed that, for most cases, a single test pattern can test a maximum of 5–10% of the nets for transition faults, and the sensitized paths (except clock logic cones) had a small amount of branching. Our analysis showed that the number of fan-out pins was three or fewer for 95% of all instances. Although we found some cases in which the number of fan-outs was high, the majority of them were on clock logic cones. In our sensitized path analysis, we excluded clock cones. As a result, the run time of the sensitized path search procedure was considerably shorter than the ATPG run time.

Our simulations showed that for the given AMD circuit blocks, there were no sensitized paths longer than T_{func}. As a result, setting T_{CRT} to T_{func} led to selection of no patterns. Thus, we do not present results for $T_{CRT} = T_{func}$.

To minimize total run time, we integrated the deviation computation procedure into our sensitized path search tool, which we named Pathfinder. As a result, Pathfinder computes the output deviations and finds all sensitized paths at the same time. In the next step, it assigns all sensitized paths a weight equal to the output deviation of their endpoints. We define the weight of a test pattern as the sum of the weights of all the paths sensitized by this pattern.

The weights of test patterns drive pattern selection. This approach selects patterns with the greatest weights first. However, it is possible that some of the sensitized paths of two different patterns are the same. If the selected patterns have already detected a path, it is not necessary to use it for evaluating the remaining patterns. The objective of this method is to minimize the number of selected patterns while still sensitizing most delay-sensitive paths.

The proposed pattern selection procedure orders the largest weighted pattern first. After selecting this pattern, we recalculate the weight of all the remaining patterns by excluding paths detected by the selected pattern. Then, this approach selects the pattern with the greatest weight in the remaining pattern set and repeats this procedure until it meets some stopping criterion (e.g., the number of selected patterns or the minimum allowed pattern weight). Since this approach sorts the selected patterns based on effectiveness, there is no need to re-sort the final set of patterns.

The final stage of the proposed method is to run top-off TDF ATPG to recover TDF coverage. In all cases, top-off ATPG generated much fewer patterns than the 1-detect TDF pattern count. Since the main purpose of this chapter is to show the application of pattern selection on industrial circuits, we do not present results for this step.

6.3.3 Comparison to SSTA-Based Techniques

The proposed method is comparable to SSTA-based techniques (e.g., [Chao 2004], [Lee 2005]). Table 6.3 illustrates the summary of this comparison, including both the original method of output deviations and the new method. Both the work of Chao et al. [Chao 2004] and the proposed work present a transition test pattern quality metric for the detection of SDDs in the presence of process variations. The focus of the work of Lee et al. [Lee 2005], on the other hand, was to present a timing hazard-aware SSTA-based technique for the same target defect group. The work of Chao et al. [Chao 2004] and the proposed work do not cover timing hazards. The formulation is different in these methods. Chao and coworkers [Chao 2004] run dynamic timing analysis multiple times for each test pattern to create a delay distribution, using simple operators (e.g., +/-) while propagating the delay values. Lee et al. [Lee 2005] run statistical dynamic timing analysis once for each test pattern, using simple operators for delay propagation (similar to [Chao 2004]), but the analysis of timing hazards adds complexity to their formulation. Both of these methods assume a Gaussian delay distribution.

On the other hand, in the original proposed work, there was no assumption regarding the shape of the delay distribution. This is because we use probability values instead of distributions. We compute the metric using probability propagation. The drawback of the original proposed method is that its effectiveness drops if the combinational depth of the circuit is very large (i.e., greater than 10). The new proposed method overcomes this limitation. Similarly, both previous studies [Chao 2004; Lee 2005] and the new proposed method can handle large combinational depths; using CLT, it could be argued that their accuracy might increase with the increased combinational depth. We expect the run time of SSTA-based WHAT [Chao 2004; Lee 2005] to limit its applicability to industrial-size designs. Further optimization might eliminate this shortcoming. On the other hand, the proposed method is quick, and its run time increases less rapidly with increase in circuit size. Since the new proposed method uses a path enumeration procedure, the run time could be a limiting factor if the number of sensitized paths is very large. However, our simulations showed that this limitation was not observed on the tested circuit blocks.

6.4 Simulation Results

In this section, we present experimental results obtained for four industrial circuit blocks. We provide details for the designs and the experimental setup in Section 6.4.1, then present simulation results in Sections 6.4.2 and 6.4.3.

TABLE 6.3

Comparison of SSTA-Based Approaches

Subject	Chao 2004	Lee 2005	Deviation Based, Original	Deviation Based, New
Main topic	Presents an SSTA-based coverage metric for estimating test quality of transition test patterns under process variation effects.	Presents a timing hazard-aware statistical timing method that can be applied to transition test patterns under process variation effects.	Presents a deviation-based coverage metric for estimating test quality of transition test patterns under process variation effects.	Presents an SSTA-based output deviations coverage metric for estimating test quality of transition test patterns under process variation effects.
Pattern selection	Proposes a pattern selection procedure based on the defined metric.	Proposes a pattern selection method based on signal slacks.	Proposes a pattern selection procedure based on the defined metric.	Proposes a pattern selection procedure based on the defined metric.
Timing hazards	Cannot handle timing hazards.	Can handle timing hazards.	Cannot handle timing hazards.	Cannot handle timing hazards.
Metric computation	Monte Carlo-based dynamic timing simulation.	Timing hazard-aware statistical dynamic timing simulation.	Probability propagation.	Propagation of path delay mean and variance.
	Requires pin-to-pin timing for each cell as a Gaussian probability density function.	Requires pin-to-pin timing for each cell as a Gaussian probability density function.	Requires pin-to-pin delay defect probability for each cell. It can handle any type of distribution because it uses the actual probability value rather than the distribution.	Requires probability density function mean and variance of pin-to-pin timing for each cell. More accurate if Gaussian probability density function is assumed.

	Formulation is simple and effective because only simple operators are needed.	Although formulation for ±/min/max operators is simple, considering hazards requires more complicated analysis for each test pattern.	Formulation is more complicated than that of Chao et al. [Chao 2004] due to probability propagation but is simpler than that of Lee et al. [Lee 2005].	Formulation is simple and effective since only simple operators are needed.
	Requires running dynamic timing analysis multiple times for each pattern; hence, the expected run time is longer.	Requires running statistical dynamic timing analysis once for each pattern, but timing hazard analysis increases run time considerably.	Requires a single pass of probability propagation for each pattern; hence, run time is short.	Requires a single pass of mean/variance propagation for each pattern, followed by path enumeration. Run time depends on the number of sensitized paths per pattern and can be long, but is shown to be shorter than the ATPG time.
Applicability	It can handle very large combinational depths.	It can handle very large combinational depths.	The effectiveness of the method decreases for very large combinational depths due to saturation of defect probabilities along the sensitized paths.	It can handle very large combinational depths.
	Application to industrial designs might require further optimization due to long run time.	Application to industrial designs might require further optimization due to long run time.	Application to industrial circuits is feasible because the run time is linear with circuit size. However, practical aspects might be a limiting factor, as discussed in Section 6.3.2.	Application to industrial circuits is feasible.

TABLE 6.4

Description of the Circuit Blocks

Benchmark	Functionality
Circuit A	Cache block
Circuit B	Execution unit
Circuit C	Execution unit
Circuit D	Load-store unit

6.4.1 Experimental Setup and Benchmarks

We performed all experiments on a pool of state-of-the-art servers running Linux with at least four processors available at all times and 16 GB of memory. We used Pathfinder to compute output deviations and find sensitized paths and implemented select patterns using C++. We used a commercial ATPG tool to generate n-detect TDF test patterns and timing-aware TDF patterns for these circuits. We forced the ATPG tool to generate launch-off-capture (LOC) transition fault patterns. We prevented the primary input change during capture cycles and the observation of primary outputs to simulate realistic test environments. We performed all ATPG runs and other simulations in parallel on four processors.

We selected blocks from functional units of state-of-the-art AMD microprocessor designs. Each block has a different functionality. Table 6.4 shows the functionality of each block.

While generating n-detect ($n = 5$ and $n = 8$) TDF patterns, we placed no limits on the number of patterns generated and set no target for fault coverage. We allowed the tool to run in automatic optimization mode. In this mode, the ATPG tool sets compaction and ATPG efforts, determines ATPG abort limits, and controls similar user-controlled options. While generating timing-aware ATPG patterns, we used two different optimization modes. In ta-1 mode, we forced the tool to sensitize the longest paths to obtain the highest-quality test patterns, at the expense of increased CPU run time and pattern count. In ta-2 mode, we relaxed the optimization criteria to decrease run time and pattern count penalty.

We obtained timing information for gate instances from a timing library, as described in Section 6.3.2. Both the ATPG and Pathfinder tools used the same timing data. The ATPG tool used nominal delay values because it was not designed to use delay variations. We did not model interconnect delays, leaving that for future work. We allowed the pattern selection in Pathfinder to continue until it selected all nonzero weight patterns.

6.4.2 Simulation Results

We grouped the simulations for generating patterns into three main categories:

1. **n-detect TDF ATPG:** We generated patterns for a range of multiple-detect values. We used $n = 1$, 3, 5, and 8. The results for $n = 5$ and

$n = 8$ are shown in this section. We used the n-detect TDF pattern set because we needed a pattern repository that was likely to sensitize many long paths, comparable to the number of long paths sensitized by timing-aware ATPG. Using a large number of patterns and n-detect TDF ATPG satisfied this requirement.

2. **Timing-aware ATPG using different optimization modes:** We generated timing-aware patterns for optimization modes ta-1 and ta-2, as described previously.

3. **Selected patterns:** We used our in-house Pathfinder tool to select high-quality patterns from both n-detect and timing-aware pattern sets.

We observed the pattern count for timing-aware ATPG was much greater than the TDF ATPG pattern count. As n increased, the number of patterns in the n-detect pattern set also increased. Figure 6.6 shows the number of test patterns generated by n-detect ATPG and timing-aware ATPG and the number of patterns selected by the proposed method (while retaining the long-path coverage provided by the much larger test set). Figure 6.6 shows

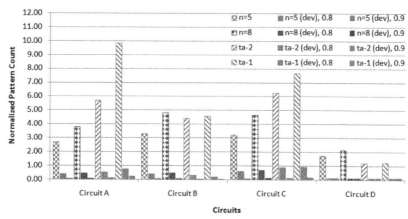

n=5	5-detect TDF pattern set
n=5 (dev), 0.8	Selected patterns from 5-detect TDF pattern set when $T_{CRT}=0.8\ T_{func}$
n=5 (dev), 0.9	Selected patterns from 5-detect TDF pattern set when $T_{CRT}=0.9\ T_{func}$
n=8	8-detect TDF pattern set
n=8 (dev), 0.8	Selected patterns from 8-detect TDF pattern set when $T_{CRT}=0.8\ T_{func}$
n=8 (dev), 0.9	Selected patterns from 8-detect TDF pattern set when $T_{CRT}=0.9\ T_{func}$
ta-2	Timing-aware pattern set with optimization mode ta-2
ta-2 (dev), 0.8	Selected patterns from ta-2 when $T_{CRT}=0.8\ T_{func}$
ta-2 (dev), 0.9	Selected patterns from ta-2 when $T_{CRT}=0.9\ T_{func}$
ta-1	Timing-aware pattern set with optimization mode ta-1
ta-1 (dev), 0.8	Selected patterns from ta-1 when $T_{CRT}=0.8\ T_{func}$
ta-1 (dev), 0.9	Selected patterns from ta-1 when $T_{CRT}=0.9\ T_{func}$

FIGURE 6.6

Normalized number of test patterns for n-detect ATPG, timing-aware ATPG (ta-1 and ta-2), and the proposed pattern selection method (dev). T_{CRT} was set to $0.8 * T_{func}$ and $0.9 * T_{func}$ for the proposed method. All values were normalized by the result for $n - 1$.

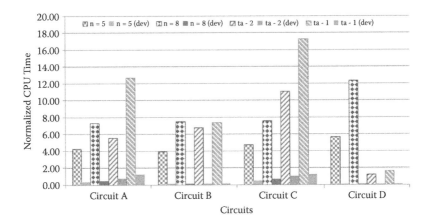

FIGURE 6.7
Normalized CPU time usage for n-detect ATPG, timing-aware ATPG (ta-1 and ta-2), and the proposed pattern-grading and pattern selection method (dev). T_{CRT} was set to $0.8 * T_{func}$ for the proposed method. All values were normalized by the result for $n = 1$.

results for $T_{CRT} = 0.8 * T_{func}$ and $T_{CRT} = 0.9 * T_{func}$. We find that, for all cases, the number of patterns selected by the proposed method was only a very small fraction of the overall pattern set. When $T_{CRT} = 0.8 * T_{func}$, the proposed scheme selected only 7% of the available patterns for circuit A from the timing-aware pattern set ta-1. We obtained similar results for other benchmarks. As expected, as T_{CRT} increased, the number of selected patterns dropped as low as 3% of the original pattern set.

The results for CPU runtime usage were as striking as the pattern count results. Figure 6.7 shows the normalized CPU runtime usage results for n-detect ATPG, timing-aware ATPG (ta-1 and ta-2), and the proposed pattern grading and pattern selection method (dev). As seen, the complete processing time (pattern grading and pattern selection) for the proposed scheme was only a fraction of the ATPG run time. For instance, for circuit C, $n = 8$, ATPG run time was 10 times longer than the pattern grading and selection time. For circuit B, the time spent for pattern grading and selection was only 2.5% of the ta-1 timing-aware ATPG run time. Note that the CPU run time of the proposed method depended on the applied T_{CRT}. Very low T_{CRT} values could increase the run time because the number of paths discarded on the fly will be lower.

Since the proposed scheme allowed us to select as many patterns as needed to cover all high-risk paths, the patterns selected by the proposed scheme sensitized all of the long paths that can be excited by the given base pattern set using only a fraction of the test patterns. Note that the effectiveness of the base pattern set bound the effectiveness of our method.

The long-path coverage ramp up for the selected patterns was also significantly better than both n-detect and timing-aware ATPG patterns. Figure 6.8

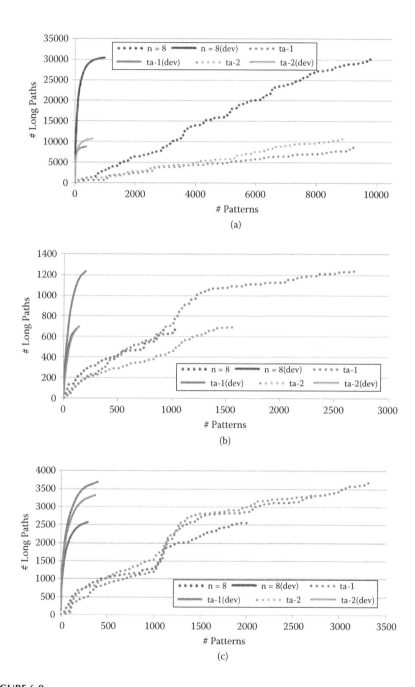

FIGURE 6.8
The long-path coverage ramp up using the base 8-detect TDF ATPG ($n = 8$), timing-aware ATPG in modes ta-1 and ta-2, selected patterns from 8-detect TDF ATPG [$n = 8$ (dev)] and timing-aware ATPG [ta-1 (dev), ta-2 (dev)]. T_{CRT} was 0.8 * T_{func} for the proposed method: (a) circuit A, (b) circuit D, and (c) circuit C.

presents the results for the long-path coverage ramp up with respect to the number of applied patterns. For all cases, the selected and sorted patterns covered the same number of long paths much faster and used far fewer patterns.

6.4.3 Comparison of the Original Method to the Modified Method

In this section, we compare the original deviation-based method [Yilmaz 2008c] to the new method proposed. We used three ASIC-like IWLS benchmark circuits for this comparison. Note that, although the Register Transfer Level (RTL) codes of these benchmarks were the same as the ones used by Yilmaz and coworkers [Yilmaz 2008c], the synthesized net lists were different due to library and optimization differences. To make a fair comparison, we reimplemented the original method [Yilmaz 2008c] to use the same pattern selection method proposed in this chapter. The only difference between those methods was this new procedure to calculate metrics. We ran all simulations on servers with similar configurations.

Figure 6.9 shows a comparison of CPU run time between the original and the modified method. As seen, for all cases, the new method had a superior CPU run time compared with the original method. The main reason for the difference between CPU run times is that the original method consumed more time to evaluate patterns due to the larger number of selected patterns. Depending on the benchmark, the impact of this effect on the overall CPU run time might be considerable, as in the case of aes_core, or it might be negligible, as in the case of systemcaes.

Figure 6.10 presents the normalized number of sensitized long paths for the original and the modified methods. Although the modified method consistently sensitized more long paths than the original method, the difference was rather small. We can better analyze Figure 6.10 if we consider

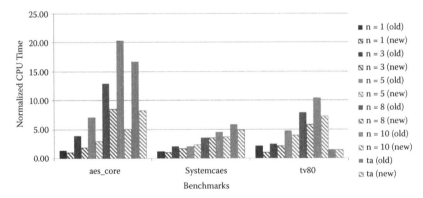

FIGURE 6.9
Normalized CPU run time for the selected n-detect ATPG and timing-aware ATPG patterns for the original method (old) and the modified method (new). All values were normalized by the result for $n = 1$ (new).

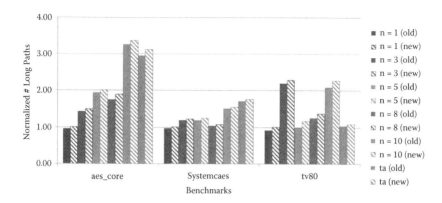

FIGURE 6.10
Normalized number of sensitized long paths for the selected n-detect ATPG and timing-aware ATPG patterns for the original method (old) and the modified method (new). All values were normalized by the result for $n = 1$ (new).

Figure 6.11. Figure 6.11 shows the normalized number of selected patterns for each method. As seen, the modified method consistently selected fewer patterns than the original method. The difference was significant for benchmarks systemcaes and tv80. Although the number of sensitized long paths was similar for these methods, the number of selected patterns was significantly different. This result shows that the modified method was more efficient than the original method in selecting high-quality patterns. The main reason for the difference in the number of selected patterns was that, due to deviation saturation on long paths, the original method was unable to distinguish between long paths and shorter paths. As a result, the original method selected more patterns to cover all of these paths.

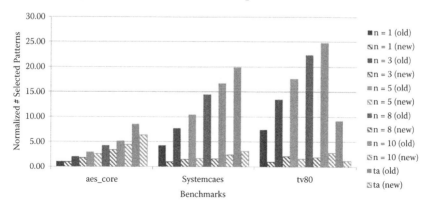

FIGURE 6.11
Normalized number of selected patterns from the base n-detect ATPG and timing-aware ATPG patterns for the original method (old) and the modified method (new). All values were normalized by the result for $n - 1$ (new).

6.5 Conclusions

We have presented a test-grading technique based on output deviations for SDDs and applied it to industrial circuits. We have redefined the concept of output deviations to apply it to industrial circuit blocks and have shown we can use it as an efficient surrogate metric to model the effectiveness of TDF patterns for SDDs. Experimental results for the industrial circuits showed the proposed method intelligently selected the best set of patterns for SDD detection from an n-detect or timing-aware TDF pattern set, and it excited the same number of long paths compared with a commercial timing-aware ATPG tool using only a fraction of the test patterns and with negligible CPU runtime overhead.

6.6 Acknowledgments

Thanks are offered to Prof. Krishnendu Chakrabarty of Duke University and Prof. Mohammad Tehranipoor and Ke Peng of the University of Connecticut for their contributions to this chapter. Colleagues at AMD are thanked for valuable discussions.

References

[Ahmed 2006b] N. Ahmed, M. Tehranipoor, and V. Jayram, Timing-based delay test for screening small delay defects, in *Proceedings IEEE Design Automation Conference*, 2006.

[Chao 2004] C.-T.M. Chao, L.-C. Wang, and K.-T. Cheng, Pattern selection for testing of deep sub-micron timing defects, in *Proceedings IEEE Design Automation and Test in Europe*, 2004, vol. 2, pp. 1060–1065.

[Cheng 2000] K.-T. Cheng, S. Dey, M. Rodgers, and K. Roy, Test challenges for deep sub-micron technologies, in *Proceedings IEEE Design Automation Conference*, 2000, pp. 142–149.

[Ferhani 2006] F.-F. Ferhani and E. McCluskey, Classifying bad chips and ordering test sets, in *Proceedings IEEE International Test Conference*, 2006.

[Forzan 2007] C. Forzan and D. Pandini, Why we need statistical static timing analysis, in *Proceedings IEEE International Conference on Computer Design*, 2007, pp. 91–96.

[Keller 2004] B. Keller, M. Tegethoff, T. Bartenstein, and V. Chickermane, An economic analysis and ROI model for nanometer test, in *Proceedings IEEE International Test Conference*, 2004, pp. 518–524.

[Lee 2005] B.N. Lee, L.-C. Wang, and M. Abadir, Reducing pattern delay variations for screening frequency dependent defects, in *Proceedings IEEE VLSI Test Symposium*, 2005, pp. 153–160.

[Nitta 2007] I. Nitta, S. Toshiyuki, and H. Katsumi, Statistical static timing analysis technology, *Journal of Fujitsu Science and Technology* 43(4), pp. 516–523, October 2007.

[Padmanaban 2004] S. Padmanaban and S. Tragoudas, A critical path selection method for delay testing, in *Proceedings IEEE International Test Conference*, 2004, pp. 232–241.

[Park 1989] E.S. Park, M.R. Mercer, and T.W. Williams, A statistical model for delay-fault testing, *IEEE Design and Test of Computers* 6(1), pp. 45–55, 1989.

[Trivedi 2001] K. Trivedi, *Probability and Statistics with Reliability, Queuing, and Computer Science Applications,* 2nd edition, Wiley, New York, 2001.

[Waicukauski 1987] J.A. Waicukauski, E. Lindbloom, B.K. Rosen, and V.S. Iyengar, Transition fault simulation, *IEEE Design and Test of Computer* 4(2), pp. 32–38, April 1987.

[Wang 2008a] Z. Wang and K. Chakrabarty, Test-quality/cost optimization using output-deviation-based reordering of test patterns, *IEEE Transactions on Computer-Aided Design of Integrated Circuits and Systems* 27, pp. 352–365, February 2008.

[Yilmaz 2008b] M. Yilmaz, K. Chakrabarty, and M. Tehranipoor, Interconnect-aware and layout-oriented test-pattern selection for small-delay defects, in *Proceedings International Test Conference [ITC'08]*, 2008.

[Yilmaz 2008c] M. Yilmaz, K. Chakrabarty, and M. Tehranipoor, Test-pattern grading and pattern selection for small-delay defects, in *Proceedings IEEE VLSI Test Symposium*, 2008, pp. 233–239.

[Yilmaz 2010] M. Yilmaz, K. Chakrabarty, and M. Tehranipoor, Test-pattern selection for screening small-delay defects in very-deep sub-micrometer integrated circuits, *IEEE Transactions on Computer-Aided Design of Integrated Circuits and Systems* 29(5), pp. 760–773, 2010.

[Zolotov 2010] V. Zolotov, J. Xiong, H. Fatemi, and C. Visweswariah, Statistical path selection for at-speed test, *IEEE Transactions on Computer-Aided Design of Integrated Circuits and Systems* 29(5), pp. 749–759, May 2010.

7

Hybrid/Top-off Test Pattern Generation Schemes for Small-Delay Defects

Sandeep K. Goel and Narendra Devta-Prasanna

CONTENTS

7.1 Introduction

Advances in design methods and process technologies are causing a continuous increase in the complexity of integrated circuits (ICs). The increased complexity and nanometer feature sizes of modern ICs make them susceptible not only to manufacturing defects but also to performance and quality issues. Process variation, power supply noise, cross talk, resistive opens/bridges, and rule violations related to design for manufacturing (DfM) such as butted contacts or insufficient via enclosures introduce small additional delays in the circuit [Mitra 2004; Kruseman 2004]. These delays, commonly referred to as small-delay defects (SDDs), can cause immediate failures if introduced on critical paths in the circuit. However, if they occur on noncritical paths, then they pose a quality risk. Due to these concerns, testing for SDDs is important in the latest technologies to achieve very high quality (very low defective parts per million, DPPM).

Commonly used fault models such as single stuck-at fault and transition delay fault (TDF) [Waicukauski 1987; Pomeranz 1999] models provide good coverage of most manufacturing defects, but they are insufficient to ensure very low DPPM associated with SDDs for high-performance chips.

While the stuck-at fault model is likely to miss delay defects, the transition fault model is capable of screening gross-delay defects but not very effective in detecting SDDs. It is because transitional fault pattern generation does not take into account the related timing information; hence, it is more likely to generate patterns that detect faults on easy or short paths. However, detection of SDDs requires fault activation and propagation through the longest path.

The path delay fault model can be used to target SDDs, but it requires enumeration of real paths in the design [Li 2000]. As the number of paths increases exponentially with design complexity, use of the path delay fault model for testing SDDs is limited in practice. The existing non-timing-aware transition fault patterns can be used to detect SDDs if applied at a clock speed higher than the system speed [Uzzaman 2006; Kruseman 2004]. Although these so-called faster-than-at-speed patterns have been effective in detecting SDDs, the chance of hazards and higher power consumption during tests due to the higher clock speed can cause unexplainable failures and unnecessary yield loss (overkill) due to scrapping of good parts.

Recently, several commercial electronic design automation (EDA) companies have launched timing-aware pattern generation (automatic test pattern generation, ATPG) tools for testing of SDDs [Lin 2006; Uzzaman 2006; Kapur 2007]. Timing-aware ATPG takes the delay information from the standard delay format (SDF) into account, and for each transition fault, it tries to excite and observe the fault effect along the longest path. Such an approach detects a wider range of delay defects while operating the circuit at the rated clock speed. However, when these ATPG tools are run in timing-aware mode, the resulting test pattern count and tool run time are increased significantly [Yilmaz 2008c]; hence, they may not be practical. Therefore, efficient methods are required that reduce the pattern count while achieving similar or higher delay test coverage (DTC) in less run time.

In this chapter, two approaches for efficient timing-aware transition fault pattern generation are presented that provide higher SDD coverage with a reasonable increase in the number of test patterns. In both approaches, a subset of transition faults is identified that should be targeted by the timing-aware ATPG; for the rest of the faults, classic non-timing-aware transition fault patterns can be generated. As only a fraction of faults is considered for timing-aware ATPG, the proposed approaches also result in runtime savings. The first approach has an advantage of higher DTC; the pattern count for the second approach is lower compared to the first method. For several industrial circuits, a comparative analysis of existing and proposed methods of generating timing-aware test patterns for SDDs is presented. The presented approaches require little or no modification in the existing commercial ATPG and achieve better or similar results compared to existing timing-aware ATPG approaches.

7.2 Fault Set for Timing-Aware ATPG

In the timing-aware mode (as described in Chapter 3), most ATPG tools try to excite and propagate delay fault effect along the longest path (also the least-slack path) for the corresponding fault. Due to inherent nature of enumerating a number of paths for each fault before generating a test pattern, timing-aware ATPG results in significantly higher run time. However, targeting all faults in the timing-aware mode is not really required as for a large number of faults (typically > 50%), non-timing-aware transition fault ATPG does a very good job already either by accidentally detecting faults along the longest path or due to only a limited number of possible testable paths for the fault. This is also the main motivation behind the approaches presented in this chapter. Validity of such assumption is basically dependent on circuit topology and slack distribution of paths in a circuit.

Figure 7.1 shows the path length distribution (worst case) for four different industrial circuits. We have selected these circuits as their path length profiles vary from wide-and-relax (circuits A and B) to narrow-and-tight (circuits C and D) distribution. Details about these circuits are described in Section 7.4. In Figure 7.1, the *x*-axis shows the path length; the *y*-axis shows the percentage of paths in the circuit. From the figure, we can see that circuits A and B have wide path distribution with slack ranging from 30% to 90%. The slack on a path is defined as the difference between the clock period and the path length. Contrary to this, circuits C and D have very narrow and tight path distribution, with slack ranging from 30% to 50% and 10% to 30%, respectively.

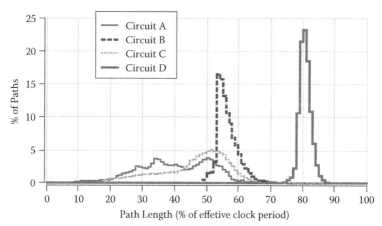

FIGURE 7.1
Path length profiles for different circuits.

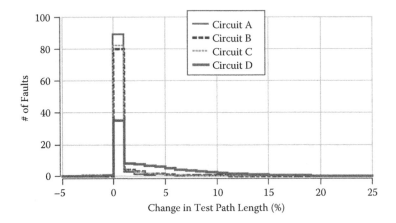

FIGURE 7.2
Change in test path length between timing-aware and non-timing-aware ATPG.

For all detected transition faults in the circuit, Figure 7.2 shows the change in test path length between non-timing-aware and timing-aware transition fault pattern generation. For test pattern generation, we used the Mentor Graphics FastScan ATPG tool [Lin 2006]. For the timing-aware ATPG, we used fault dropping based on the slack margin concept [Lin 2006] with a slack margin value of 5%, which means a fault was dropped during fault simulation if it was detected along a path with slack Ta that satisfies the following criterion: $(Ta - Ts)/Ta < 0.05$, where Ts is the slack of the longest static path for the fault. However, similar results were obtained for other slack margin limits also.

From the figure, we can see that for all circuits, there is no change in the test path length for a large number of faults. For circuit A, around 85% of the detected faults have the same test path length, while for circuit C, 80% of the faults have the same test path length. For circuit D, which has very narrow and tight path length distribution, around 38% of the total faults have no change in the path length. This means that even the traditional non-timing-aware ATPG was able to excite and propagate fault effects through the same paths as the timing-aware ATPG. This confirms our basic assumption that running timing-aware ATPG on all faults in a given circuit is not really required.

Now, the challenge is to identify faults in a circuit that do not require timing-aware ATPG. One way to handle this is to run both timing and non-timing-aware ATPG on the circuit and then exclude faults that do not see any change in their path length. For the excluded faults, a non-timing-aware run should be performed; for the remaining faults, timing-aware patterns can be generated. Although this approach would result in a low pattern count, this approach would be time consuming as multiple timing-/non-timing-aware runs are required. Therefore, a runtime efficient heuristic for filtering faults that should be targeted for timing-aware ATPG is critical.

7.3 Small-Delay Defect Pattern Generation

Two runtime efficient approaches [Goel 2009] that use a heuristic-based fault-filtering criterion are described in this section. The first approach results in higher SDD coverage; the second approach results in lower pattern count.

7.3.1 Approach 1: TDF plus Top-off SDD

Figure 7.3 shows an outline of approach 1. In this method, first we generate traditional delay fault patterns for all the faults in the given circuit (shown as TDF patterns in the figure). Please note that these patterns can be for TDFs or in-line resistive faults (IRFs). These patterns have to be generated anyway to catch gross-delay defects. Once the TDF pattern set is generated, it is fault graded for SDDs by setting the ATPG in the timing-aware fault simulation mode.

Next, we filter the timing-aware target-detected faults. Different filtering criteria can be used. We use the following criterion based on a user-defined parameter τ ($0 < \tau \leq 1$). A fault is considered to be a timing-aware target if $(Ta - Ts)/Ts \geq \tau$; otherwise, the fault is labeled as a non-timing-aware fault.

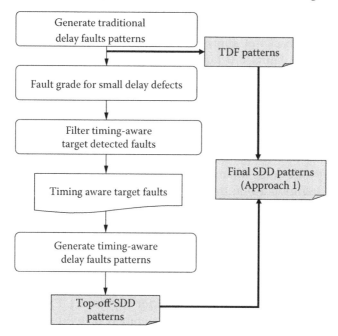

FIGURE 7.3
Flowchart for approach 1.

Ta and *Ts* are the actual test path slack and the slack on the longest static path for the fault, respectively. These two parameters are already reported by the commercial ATPG tool [Lin 2006]. The motivation here is that for a fault, if the slack of the path selected by ATPG is higher than a particular limit, then the fault should be passed for timing-aware ATPG to improve its path length. The longer the path length, the higher the DTC is. Please note that the parameter τ is similar to the slack margin parameter δ defined by Lin et al. [Lin 2006] but with a difference that we divide by maximum static slack instead of actual test slack.

Once the timing-aware target fault list is filtered, timing-aware ATPG is carried out for the faults (resulting patterns are shown as top-off SDD patterns). The combined TDF and top-off SDD pattern set represents the final SDD-targeted pattern set. We call this approach TDF plus top-off SDD as only top-off timing-aware patterns are generated in addition to the traditional TDF pattern set. This approach results in higher DTC than non-timing-aware TDF patterns set; the pattern count is similar or less than the timing-aware pattern set generated based on the entire fault list in the design. Furthermore, as the timing-aware ATPG is run only on a subset of faults, this approach is also more efficient in terms of run time compared to the default timing-aware approaches supported by ATPG tools.

7.3.2 Approach 2: Top-off SDD plus Top-off TDF

One of the drawbacks of approach 1 described in the previous section is that some of the faults are targeted during both non-timing-aware and timing-aware ATPG. Due to this, the resulting pattern set might contain redundant patterns and result in a higher pattern count than necessary. To circumvent this problem, we propose a modification to approach 1. The new approach is outlined in Figure 7.4.

The flow for this approach is similar to approach 1. Here, once the top-off SDD patterns have been generated, these patterns are fault simulated for the non-timing-aware faults filtered previously. All detected faults are removed, and a new traditional non-timing-aware ATPG run is carried out for all remaining undetected faults. The generated patterns are shown as top-off TDF patterns. The combined top-off SDD and top-off TDF pattern set represents the final SDD targeted pattern set.

We call this approach top-off SDD plus top-off TDF. The size of top-off TDF pattern set is smaller than the original TDF pattern set; therefore, this approach results in further substantial savings in pattern count. However, the DTC for this approach is lower than for approach 1. Please note that to further minimize the pattern count, generation of top-off TDF and SDD patterns can also be done in an iterative manner. However, that will result in longer run time and less delay defect coverage.

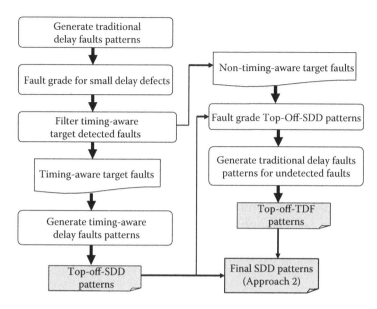

FIGURE 7.4
Flowchart for approach 2.

7.4 Experimental Results

Experimental results for six small industrial circuits are presented in this section. These six circuits were selected because they have varied path length distribution. In experiments, commercial the ATPG tool FastScan from Mentor Graphics was used. However, the described approaches are not limited to FastScan only, and any timing-aware ATPG can be used instead. Table 7.1 shows the basic test characteristics for the selected six circuits. Column 2 shows the total number of noncollapsed faults in the

TABLE 7.1

Characteristics of Benchmark Circuits

		Traditional TDF		
Circuit	Number of Faults	Patterns	TC (%)	DTC (%)
A	83,591	161	90.55	77.22
B	866,258	1,115	93.41	80.79
C	66,261	341	88.47	72.28
D	647,019	2,018	89.53	75.29
E	11,325	173	62.22	52.84
F	18,898	132	96.23	90.21

TC, test coverage; DTC, delay test coverage.

circuit; column 3 shows the total number of non-timing-aware transition fault patterns for the circuit. Column 4 and 5 show the noncollapsed test coverage and DTC, respectively, for the circuits. Please also note that we used a transition test based on launch off capture (LOC), and changes on primary inputs during capture cycles and observation of primary outputs were prevented.

First, we compare the results obtained from the proposed approaches and the default timing-aware ATPG supported in FastScan [Lin 2006] to traditional non-timing-aware ATPG. For effectiveness, we compare the pattern count and DTC metric [Lin 2006] reported by the tool. In our experiments, we used the dropping based on slack margin (DSM) concept as proposed by Lin et al. [Lin 2006] to maximize DTC. The DTC increases with a decrease in slack margin sm. For a range of slack margins sm, Table 7.2 shows test pattern count and DTC results obtained from timing-aware ATPG and the proposed two approaches. In our experiments, we considered $\tau = sm$ for fault filtering.

TABLE 7.2

Test Pattern and DTC Results for Timing-Aware ATPG and the Proposed Two Approaches

Circuit	sm	Timing-Aware TDF [Mitra 2004]		TDF + Top-off SDD (Approach 1)		Top-off SDD + Top-off TDF (Approach 2)	
		Patterns	DTC (%)	Patterns	DTC (%)	Patterns	DTC (%)
A	5	454	78.09	396	**78.11**	282	77.71
	10	469	78.04	303	77.98	204	77.56
	15	494	78.05	259	77.87	174	77.41
B	5	3,050	84.26	3,006	84.04	2,249	83.65
	10	2,826	84.07	2,559	83.53	1,882	83.03
	15	2,870	84.03	2,276	83.01	1,683	82.60
C	5	707	73.37	739	**73.51**	493	72.98
	10	767	73.44	709	73.37	489	72.81
	15	779	73.45	679	73.30	506	72.78
D	5	16,724	81.16	16,699	**81.17**	15,192	80.99
	10	14,114	80.89	14,310	**80.95**	12,741	80.72
	15	11,891	80.68	12,069	80.65	10,588	80.38
E	5	333	53.71	337	**53.82**	248	53.40
	10	305	53.60	306	**53.67**	215	53.22
	15	53%	53.31	310	**53.69**	224	53.13
F	5	210	90.78	234	**90.87**	186	90.50
	10	218	90.87	237	**90.89**	193	90.56
	15	219	90.74	205	90.70	166	90.29
Average	10	203%	2.04	184%	1.96	127%	1.55

Note: Bold entries in DTC column for approach one indicate the case when approach one resulted in better/higher DTC than other approaches in the table.

Column 2 shows the percentage of slack margin; that is, a value of 5 means $\delta = 0.05$. Columns 3 and 4 show the pattern count and DTC for timing-aware ATPG, respectively. Columns 5 and 6 show the pattern count and DTC for approach 1; columns 7 and 8 show the same for approach 2. The last row (Average) in the table shows the average percentage increase in pattern count and average absolute improvement in DTC obtained from respective approaches compared to the traditional non-timing-aware transition fault ATPG results (as shown in Table 7.1). From Table 7.2, we can see that, compared to non-timing-aware ATPG, use of timing-aware ATPG resulted in an average increase of 203% in test pattern count, while the average absolute improvement in DTC was 2.04%. Approach 1 resulted in an average 19% (203 − 184 = 19) lower pattern count than timing-aware ATPG; the average penalty in DTC was around 0.08% (2.04 − 1.96 = 0.08). As expected, approach 2 resulted in a minimum increase in the pattern count. Compared to the timing-aware ATPG, approach 2 resulted in pattern count savings up to 75% (203 − 127 = 75) at an average expense of 0.49% in DTC. For some cases, approach 1 resulted in a higher pattern count than timing-aware ATPG due to multiple detections of the same faults as mentioned in Section 7.2. However, for these cases, the DTC was also highest for approach 1. Therefore, if achieving maximum DTC is a requirement, approach 1 can be used.

Figure 7.5 shows the number of non-timing-aware TDF patterns in the final pattern set for the two proposed approaches. For approach 1, the number of non-timing-aware TDF patterns did not change with slack margin; for approach 2, they decreased with the decrease in slack margin. This is to be expected as more timing-aware patterns are generated for lower slack margins; hence, more faults are dropped during top-off SDD fault simulation.

In volume production, test pattern truncation is carried out as the entire timing-aware pattern set cannot be loaded on the tester due to memory limitation. For such cases, a maximum increment in DTC with a minimum

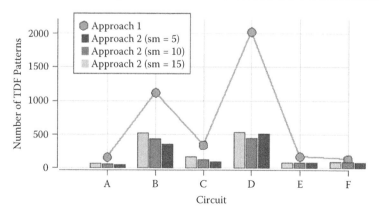

FIGURE 7.5

Non-timing-aware TDF pattern count distribution for the proposed approaches.

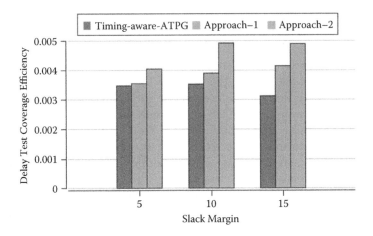

FIGURE 7.6
Comparison of delay test coverage efficiency.

increase in pattern count without sacrificing the transition fault coverage (TC) is desired. Figure 7.6 shows the DTC efficiency comparison between proposed methods and the timing-aware ATPG. DTC efficiency is defined as the average improvement in DTC per additional test pattern. From the figure, it is clear that approach 2 resulted in maximum DTC efficiency, which is expected as it also resulted in a minimum increase in pattern count.

To compare our method's effectiveness for the same number of patterns as the default timing-aware ATPG, we truncated the default timing-aware pattern set to match the pattern count obtained from approach 2. Figure 7.7 shows the difference in transition TC and DTC obtained from the truncated

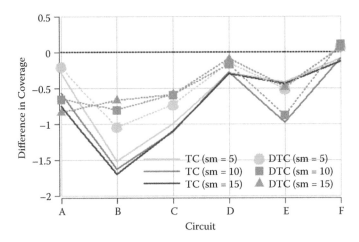

FIGURE 7.7
Difference in TC and DTC for truncated timing-aware pattern set and approach 2.

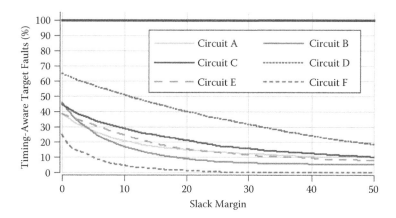

FIGURE 7.8
Target fault list for timing-aware ATPG.

pattern set and approach 2. From the figure, we can see that approach 2 outperformed the truncated timing-aware pattern set. The truncated timing-aware pattern set resulted in 0.13% to 1.62% lower TC and up to 1.04% lower DTC. Therefore, when the automatic test equipment (ATE) vector memory depth is a hard constraint, use of approach 2 is recommended.

Next, we compare runtime results for the proposed approaches with timing-aware ATPG. The total number of timing-aware target faults has a significant impact on the run time of the proposed approaches. For a range of slack margins, Figure 7.8 shows the percentage of total faults that were targeted for the timing-aware ATPG run in our proposed approaches.

It is clear from the figure that for lower slack margins, a higher number of faults needs to be considered for the timing-aware run. For very high slack margins, all faults are dropped, and no timing-aware run is required. Ideally, for a given slack margin, we would like to exclude all faults (around 80% as shown in Figure 7.2) that will not have any change in the path length. However, the filtering criterion used in our experiments seemed rather relaxed as 25% to 68% of the faults still needed to be considered for the timing-aware run. Please note that for default timing-aware ATPG, irrespective of slack margin, 100% of the faults need to be considered for the timing-aware run.

Figure 7.9 shows the average runtime savings obtained from the proposed approaches. Compared to default timing-aware ATPG, approach 1 resulted in runtime savings ranging from 2% to 16%. For approach 2, runtime savings were lower as an additional non-timing-aware ATPG was carried out in the last step (as shown in Figure 7.4). Also, note that the runtime savings shown here are pessimistic as they include the time required to calculate the DTC (final fault simulation) for the pattern set obtained by the proposed approaches.

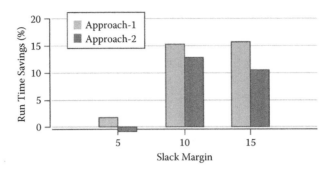

FIGURE 7.9
Runtime savings using the proposed methods.

7.5 Conclusion

Testing for SDDs is necessary to meet the high-quality and low-DPPM requirements of high-end semiconductor devices. This chapter described two approaches for effective and efficient test pattern generation targeting SDDs. The proposed methods outperformed the default timing-aware test pattern generation method supported by the commercial ATPG tools. The basic concept behind the proposed approaches is that not all faults in a circuit need to be considered for expensive timing-aware ATPG. Experimental results for several industrial circuits showed that an average 19% to 75% pattern count saving can be obtained using the proposed approaches. The average penalty in absolute DTC ranged from 0.08% to 0.49%. For several cases, the proposed approaches also resulted in the highest DTC. The proposed approaches resulted in up to 15% runtime savings. Furthermore, for most practical cases, for which the test pattern set needs to be truncated, the proposed approaches resulted in a maximum increase in DTC with a minimum increase in pattern count while maintaining the maximum transition TC.

References

[Goel 2009] S.K. Goel, N. Devta-Prasanna, and R. Turakhia, Effective and efficient test pattern generation for small delay defects, in *Proceedings IEEE VLSI Test Symposium*, 2009.

[Kapur 2007] R. Kapur, J. Zejda, and T.W. Williams, Fundamental of timing information for test: How simple can we get? In *Proceedings IEEE International Test Conference*, 2007.

[Kruseman 2004] B. Kruseman, A.K. Majhi, G. Gronthoud, and S. Eichenberger, On hazard-free patterns for fine-delay fault testing, in *Proceedings IEEE International Test Conference*, 2004.

[Li 2000] H. Li, Z. Li, and Y. Min, Reduction in number of path to be tested in delay testing, *Journal of Electronic Testing: Theory and Application [JETTA]* 16(5), pp. 477–485, 2000.

[Lin 2006] X. Lin, K.-H. Tsai, C. Wang, M. Kassab, J. Rajski, T. Kobayashi, R. Klingenberg, Y. Sato, S. Hamada, and T. Aikyo, Timing-aware ATPG for high quality at-speed testing of small delay defects, in *Proceedings IEEE Asian Test Symposium*, 2006, pp. 139–146.

[Mitra 2004] S. Mitra, E. Volkerink, E.J. McCluskey, and S. Eichenberger, Delay defect screening using process monitor structures, in *Proceedings IEEE VLSI Test Symposium*, 2004.

[Pomeranz 1999] I. Pomeranz and S.M. Reddy, On n-detection test sets and variable n-detection test sets for transition faults, in *Proceedings IEEE VLSI Test Symposium*, 1999, pp. 173–180.

[Uzzaman 2006] A. Uzzaman, M. Tegethoff, B. Li, K. McCauley, S. Hamada, and Y. Sato, Not all delay tests are the same—SDQL model shows true-time, in *Proceedings IEEE Asian Test Symposium*, 2006, pp. 147–152.

[Waicukauski 1987] J.A. Waicukauski, E. Lindbloom, B.K. Rosen, and V.S. Iyengar, Transition fault simulation, *IEEE Design and Test of Computer*, 4(2), pp. 32–38, April 1987.

[Yilmaz 2008c] M. Yilmaz, K. Chakrabarty, and M. Tehranipoor, Test-pattern grading and pattern selection for small-delay defects, in *Proceedings IEEE VLSI Test Symposium*, 2008, pp. 233–239.

8

Circuit Topology-Based Test Pattern Generation for Small-Delay Defects

Sandeep K. Goel and Krish Chakrabarty

CONTENTS

8.1 Introduction

Advances in design methods and process technology are continuing to push the envelope for integrated circuits. The use of advanced process technology brings forward several design and test challenges. In addition to manufacturing defects such as resistive opens/bridges, design-related issues such as process variation, power supply noise, cross talk, and design-for-manufacturability (DfM) rule violations such as butted contacts or insufficient via enclosures introduce small additional delays in the circuit [Mitra 2004; Kruseman 2004]. These delays are commonly referred to as small-delay defects (SDDs), and testing them is one of the major challenges that the semiconductor industry is facing today. SDDs can cause immediate failure of a circuit if introduced on critical paths, whereas they cause major quality concerns if they occur on noncritical paths. For very-high-quality products (0–100 defective parts per million [DPPM]), testing SDDs is necessary.

Conventional test methods such as the stuck-at test and transition delay fault (TDF) test [Waicukauski 1987] still provide very high coverage of most manufacturing defects but cannot guarantee sufficient coverage of SDDs. Stuck-at-based tests do not target delay defects; the traditional transition fault tests target gross-delay defects, and they are likely to miss SDDs. Path-delay-based tests can provide good coverage of SDDs; however, generation of path delay patterns suffers from practical problems, such as enumeration of long paths and test generation complexity with increasing numbers of paths.

Testing SDDs requires fault activation and propagation through long paths in the design. Several techniques [Kruseman 2004; Lin 2006, 2007a; Yilmaz 2008c; Putman 2006; Sato 2005b; Uzzaman 2006; Kapur 2007; Li 2000; Ahmed 2006b; Turakhia 2007; Yan 2004; Goel 2009; Devta-Prasanna 2009] have been proposed in the literature for testing SDDs, and they can be classified into three basic categories: (1) faster-than-at-speed, (2) timing-aware automatic test pattern generation (ATPG), and (3) pattern/fault selection-based hybrid techniques. In faster-than-at-speed techniques, existing or modified TDF patterns are applied at a clock speed higher than the system speed. Applying patterns at a faster clock speed than the rated system speed reduces the available slack in the design and helps detect SDDs. Although faster-than-at-speed methods provide very good coverage for SDDs, they suffer from several drawbacks, such as higher power consumption during a test, complex test application, and chances of hazards that can cause unexplainable failures.

Timing-aware ATPG techniques for testing SDDs are preferred among electronic design automation (EDA) companies [Lin 2006; Uzzaman 2006; Kapur 2007]. In timing-aware ATPG, delay information for every node is taken into account while generating test patterns. In these techniques, for each fault, the ATPG tool tries to excite and propagate fault effects along the longest testable path possible. Timing-aware patterns do not require test application at higher-than-system speed, although if applied can also detect some reliability defects. Some of the major drawbacks of timing-aware ATPG are a huge pattern count and long computing time [Yilmaz 2008c; Goel 2009]. Due to these problems, timing-aware ATPG is not suitable for large industrial designs.

Effective pattern selection based on metrics such as output deviation has been proposed [Yilmaz 2008a]. Here, instead of generating timing-aware patterns, effective patterns are selected from a pool of patterns [Pomeranz 1999]. Unfortunately, the quality of the results obtained from this method depends on the quality of the pool of patterns, which is hard to determine prior to pattern selection. It was shown [Goel 2009] that traditional transition fault patterns already provide good SDD coverage for a large number of faults in the circuit, and for only a fraction of the total number of nodes should timing-aware ATPG be carried out. However, the two techniques presented [Goel 2009] also fell short in terms of achieving low pattern count for large industrial designs. Therefore, efficient methods are required that provide high coverage of SDDs with a minimum increase in pattern count while maintaining maximum coverage for gross-delay defects. This is also required when the entire timing-aware pattern set cannot be loaded on the tester due to tester memory limitation, and test pattern truncation needs to be carried out.

In this chapter, circuit topology-based test pattern generation for SDDs is presented. Similar to the two approaches of Goel et al. [Goel 2009] (as described in Chapter 7), the proposed approach identifies a subset of faults for which timing-aware ATPG should be carried out; for the rest of the nodes, traditional transition fault patterns can be generated. The identification of a subset of faults is done based on the fan-out count of nodes in the design. Nodes with fan-outs higher than a certain threshold are selected. The main motivation behind this selection is that an SDD on a high fan-out node is more critical as its fault effect can be propagated through several paths and cause circuit failure. To maximize the coverage of SDDs through multiple long paths instead of one long path, all fan-out nodes for the selected node are also selected as timing-aware ATPG candidates. For several International Workshop on Logic and Synthesis (IWLS) benchmarks [IWLS 2005], detailed comparative analysis of existing methods and a circuit topology-based method of generating timing-aware test patterns for SDDs are presented. The circuit topology-based method leads to only a small increase in the pattern count and provides high coverage of gross-delay defects and SDDs. Because only a fraction of faults is considered for timing-aware ATPG, the circuit topology-based SDD pattern generation method also results in runtime savings.

8.2 Circuit Topology-Based Fault Selection

Although timing-aware ATPG is the preferred solution for SDD testing due to easy test pattern generation and application flow, it has not been adopted widely due to a large pattern count and long computing time. In the timing-aware mode, most ATPG tools try to excite and propagate a delay fault effect along the longest path (also the least-slack path) for the corresponding fault. Due to the inherent problem of enumerating a large number of paths for each fault before generating a test pattern, timing-aware ATPG results in significantly longer run time.

It has been shown [Goel 2009] (Chapter 7) that not all faults should be considered for timing-aware ATPG. According to the results shown by Goel et al. [Goel 2009], there is no change in the path length for a majority of faults (up to 80%) even if timing-aware ATPG is carried out for them. For these faults, the ATPG tool is able to propagate a fault effect through the longest path fortuitously or because there are a limited number of paths. Therefore, performing timing aware ATPG on the complete circuit is not necessary.

The challenge here is to identify a minimum subset of faults in a circuit that requires timing-aware ATPG. The fault set should not be too large because that will lead to a large pattern count and extended run time. Also, it should not be too small; otherwise, the improvement in SDD quality will be small and difficult to measure. In the work of Goel and coworkers [Goel 2009], a subset of faults was selected based on the difference in tested path length and the longest path possible for each fault. The tested and longest path lengths for each fault were obtained by performing timing-aware simulation of traditional transition fault patterns. Therefore, the fault selection was heavily dependent on the results reported by the ATPG tool. In the circuit topology-based method [Goel 2010], we look at the faults from a circuit topology point of view, and no ATPG is required to select the fault list.

In a given circuit, nodes with multiple fan-outs are potential targets for timing-aware ATPG. This is because if a defect occurs at any of these nodes, the fault effect can be propagated through multiple paths (related to fan-outs) and can affect the circuit functionality. However, the non-timing-aware ATPG may select the easiest path for these nodes, and SDDs may go undetected. To understand this, consider the two example nodes shown in Figure 8.1. For node A, there is only one propagation path possible because it does not have any fan-out. Therefore, if ATPG is able to generate a test for it, it will be through path p1, which is also the longest and only path for this node. Therefore, whether we do timing-aware ATPG or non-timing-aware ATPG, coverage of SDDs for this node will not change. Hence, this node can be excluded from the timing-aware ATPG target fault list.

Next, we look at node B; as shown in Figure 8.1b, there are two paths possible because it has a fan-out of two. Path p1 is longer than path p2. Therefore, for these types of nodes, performing timing-aware ATPG can improve the SDD coverage. Furthermore, if these nodes have already been tested for SDDs, other nodes in the circuit will also have higher coverage for SDDs if they use paths through these nodes for transition fault pattern generation.

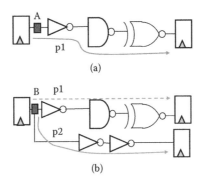

FIGURE 8.1
Examples of nodes with different fan-outs.

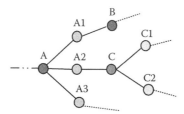

FIGURE 8.2
Fan-out node selection for multiple long-path coverage.

To further improve the coverage of SDDs, instead of selecting one long path for each fault, multiple long paths can be selected. This can be done easily by including all fan-out nodes in the list of timing-aware target faults. Consider the example shown in Figure 8.2. Node A has three fan-outs; therefore, we select nodes A, A1, A2, and A3 when we consider node A. Similarly, for node C, both C1 and C2 nodes are also considered as timing-aware target faults. If all fan-out branches are tested through their longest paths, the fan-out stem will be tested through multiple long paths.

Figure 8.3 shows the fan-out distribution of nodes for various IWLS'05 benchmark circuits [IWLS 2005]. As expected, the number of nodes decreases with an increase in the fan-out count. There are very few nodes with a fan-out count of six or higher. Therefore, to avoid a very small fault subset when selecting faults as timing-aware candidates, the fan-out count should be limited to four or five only and not more.

Figure 8.4 shows the variation in the resulting timing-aware ATPG target fault list with fan-out count for five large IWLS benchmarks. Details about these circuits can be found in Section 8.4. Figure 8.4 shows that if only nodes with two or higher fan-outs (FC2) along with their fan-out nodes are selected, then only 35% to 45% of the total faults need to be considered for computationally expensive timing-aware ATPG. For nodes with three or

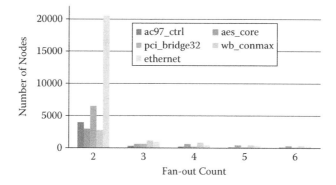

FIGURE 8.3
Nodes' fan-out profile for IWLS benchmarks.

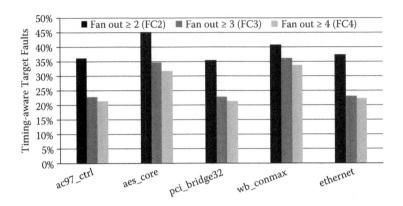

FIGURE 8.4
Variation in timing-aware ATPG target fault list with fan-out count.

more fan-outs (FC3), the total number of timing-aware ATPG faults ranges from 23% to 36%, while for four or higher fan-out nodes, it ranges from 21% to 34%. Therefore, if we consider only nodes with a certain number or more fan-outs for timing-aware ATPG, a significant reduction in run time can be obtained.

Extraction of fan-out count for nodes in a circuit is easy and does not require any functional or timing simulation. Therefore, the runtime impact of calculating fan-outs can be ignored. Next, we describe the ATPG flow that uses the fan-out-based fault selection criterion as presented in this section. Because the fan-out information is used in the proposed method for deciding which faults should be targeted for timing-aware ATPG, we also refer to this methodology as the fan-out-aware method.

8.3 SDD Pattern Generation

The fan-out-aware SDD ATPG flow is shown in Figure 8.5. The proposed flow is a hybrid one, and it requires both timing-aware and non-timing-aware transition fault ATPG. First, fan-out counts for all nodes in the design are determined. Next, based on a user-defined limit *fc*, nodes with fan-out count equal to or higher than *fc* are selected for timing-aware ATPG. For each selected node, all the corresponding fan-out nodes are also selected for timing-aware ATPG (as shown in Figure 8.2). Timing-aware ATPG is then performed on the selected nodes, and fan-out-aware SDD patterns are generated.

Next, these patterns are fault simulated for the non-timing-aware target faults filtered previously. Note that this is non-timing-aware fault simulation. All transition faults detected during the fault simulation step are marked

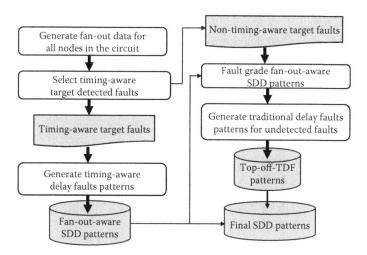

FIGURE 8.5
Fan-out-aware SDD pattern generation flow.

protected, and traditional transition fault patterns are generated only for the undetected faults. The combined pattern set for this methodology includes the fan-out-aware SDD patterns and the top-off TDF patterns. To calculate the overall coverage from the resulting pattern set, the entire pattern set is fault simulated in the timing-aware mode over all faults. During fault simulation, redundant patterns are also dropped.

For further reduction in test pattern count, several variations of the proposed ATPG flow are possible. For example, the TDF pattern can be generated first and fault simulated in timing-aware mode for timing-aware target faults. Next, only for undetected timing-aware target faults, timing-aware patterns can be generated, followed by non-timing-aware pattern generation for the remaining undetected faults. However, a complete discussion of all possible variations of the proposed flow is not considered here.

8.4 Experimental Results and Analysis

This section presents detailed experimental results for several IWLS benchmarks to show the effectiveness of the proposed method. We compare the test quality and pattern effectiveness based on several different metrics: (1) delay test coverage (DTC) [Lin 2006] as reported by ATPG, (2) total number of long paths excited by the pattern set, (3) length of the longest path excited by the pattern set, and (4) number of unique SDD locations covered by the pattern set. We also discuss the DPPM impact of the proposed method in terms of detecting randomly injected SDDs.

TABLE 8.1

Characteristics of IWLS Benchmarks

Circuit	Gate Count	Total Faults	Patterns	TDF TC (%)	DTC (%)
wb_dma	7,619	68,084	173	87.30	80.31
tv80	13,326	59,172	823	97.15	77.46
systemcaes	17,817	94,056	357	91.8	84.94
mem_ctrl	22,015	98,998	564	94.92	85.95
usb_funct	25,531	155,904	288	95.97	87.74
ac97_ctry	28,093	175,070	249	95.45	90.26
aes_cote	29,186	165,368	670	95.50	87.16
dma	41,026	165,556	656	95.48	87.32
pci_bridge32	43,907	309,216	437	93.61	86.08
wb_conmax	59,484	347,300	593	91.11	84.94
ethernet	153,948	868,248	2,805	97.31	90.42

In our experiments, we used a commercial ATPG tool; however, any other ATPG tool that supports timing-aware pattern generation can be used. Also note that, for timing-aware pattern generation, we used fault dropping based on the slack margin concept [Lin 2006] with a slack margin value of 5%, which means a fault was dropped during fault simulation if it was detected along a path with slack Ta that satisfies the following criterion: $(Ta - Ts)/Ta < 0.05$, where Ts is the slack of the longest static path for the fault. However, similar results were also obtained for other slack margin limits.

Table 8.1 shows the basic test characteristics for the selected IWLS benchmarks. Please note that all these circuits were physically synthesized and place-and-routed to obtain realistic delay values. Column 1 shows the circuit name; columns 2 and 3 show the total gate count and total number of faults, respectively. Columns 4, 5, and 6 show the number of test patterns, transition fault coverage (TC), and DTC for the traditional non-timing-aware TDF patterns, respectively.

Note that we have used a transition test based on launch off capture (LOC), and changes on primary inputs during capture cycles and observation of primary outputs were prevented. From the table, we can see that traditional TDF patterns indeed resulted in high coverage of SDDs as measured by the DTC metric.

8.4.1 Delay Test Coverage

First, we show the variation in the test pattern count and DTC for the proposed method with the number of fan-outs used during timing-aware target fault selection. For four large IWLS circuits, Figure 8.6 shows the overall pattern count obtained from the fan-out-aware method with fan-out counts 2, 3, and 4. From the figure, we can see that overall test pattern count decreased with an increase in fan-out count. This is to be expected as the

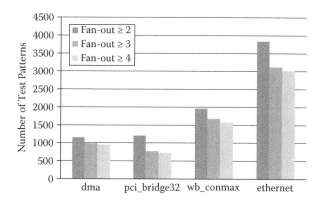

FIGURE 8.6
Pattern count variation with fan-out count.

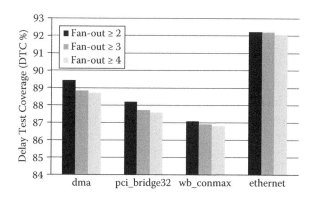

FIGURE 8.7
Variation in DTC with fan-out count.

number of timing-aware target faults decreases with an increase in fan-out count (Figure 8.4), resulting in fewer timing-aware patterns. Figure 8.7 shows the variation in overall DTC with fan-out count. Again, the DTC decreased with increase in fan-out count due to fewer timing-aware patterns, as mentioned previously.

Next, we compare the ATPG results obtained from the proposed method to the default timing-aware ATPG on the entire circuit (TA-SDD), and the two fault-filtering-based approaches presented by Goel et al. [Goel 2009]. As the approaches presented by Goel et al. [Goel 2009] outperformed the timing-critical-path-based fault-filtering approach [Lin 2007a], comparison with timing-critical-path-based fault filtering is not necessary. Table 8.2 shows the test pattern and DTC results for these approaches. Columns 2 and 3 show the pattern count and DTC for TA-SDD, respectively. Columns 4, 5, 6, and 7 show the same but for the approaches TDF plus top-off SDD (approach 1) and top-off SDD plus top-off TDF (approach 2) as described by

TABLE 8.2

Test Pattern and DTC Results

| | | | [16] | | | | Our Method | |
| | TA-SDD | | Approach 1 | | Approach 2 | | Fan-out ≥ 2 | |
Circuit	Pattern	DTC	Pattern	DTC	Pattern	DTC	Pattern	DTC
wb_dma	319	82.12	409	82.11	286	81.78	254	81.61
tv80	1,746	82.17	2,444	83.03	1,751	82.34	1,307	80.74
systemcaes	1,192	87.18	1,166	87.15	870	86.74	720	86.42
mem_ctrl	1,108	89.07	1,512	89.34	1,116	88.76	832	88.25
usb_funct	981	90.79	616	90.39	487	90.04	653	90.1
ac97_ctry	825	91.87	481	91.31	382	91.07	501	91.37
aes_cote	1,646	90.14	1,565	90.25	1,048	89.03	1,131	89.33
dma	1,625	90.16	1,518	90.29	1,032	89.02	1,144	89.42
pci_bridge32	2,147	88.84	1,390	88.62	1,132	88.3	1,199	88.21
wb_conmax	2,379	87.77	1,995	87.67	1,598	87.23	1,964	87.08
ethernet	8,726	93.56	9,342	93.44	7,188	93.14	3,481	92.25
Average	245%	2.9	246%	2.8	165%	2.3	73%	2.0

Goel et al. [Goel 2009]. Columns 8 and 9 show the pattern count and DTC for the proposed method with a fan-out count of two or more.

From the table, we can see that, for almost all circuits, the proposed method resulted in the lowest pattern count. The DTC values obtained from the proposed methods are also similar (except circuit tv80) to those obtained from other methods. The bottom row (average) in the table shows the weighted average (based on gate count) percentage increase in pattern count and weighted average absolute improvement in DTC obtained from respective approaches compared to the traditional non-timing-aware TDF (Table 8.1). Note that we took the weighted average to accurately reflect the benefits obtained for larger circuits.

From the average results, we can see that default timing-aware ATPG (TA-SDD) resulted in a 245% average increase in pattern count compared to TDF patterns, while the average absolute improvement in DTC was 2.9. Approach 1, as presented by Goel and coworkers [Goel 2009], performed worse than the TA-SDD pattern with a 246% increase in pattern count and only a 2.8 increase in DTC. Approach 2 resulted in a 165% increase in pattern count with an average increase of 2.3 in DTC. The proposed method outperformed other methods and resulted in only a 73% increase in pattern count compared to TDF, with an average increase of 2 in DTC. Compared to TA-SDD, it showed a saving of 172% in pattern count, and the loss in DTC was only 0.9.

As mentioned in Section 8.1, one of the major hurdles in industry-wide adaptation of timing-aware patterns is the huge pattern count. Industry is looking for solutions that give a maximum increment in delay coverage with

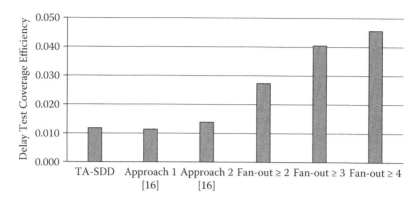

FIGURE 8.8
Delay test coverage efficiency comparison.

a minimum increase in pattern count without any loss of transition test coverage (gross-delay defects). This is also required when the entire timing-aware pattern set cannot be loaded on the tester due to tester memory limitation, and test pattern truncation needs to be carried out. To measure the effectiveness of a method in such cases, a DTC efficiency metric was presented [Goel 2009]. Delay test coverage efficiency (DTCE) is defined as the average improvement in DTC per 1% increase in overall pattern count compared to the traditional TDF pattern set. This is an important metric because what we really want is a method that can improve coverage beyond what TDF patterns have already achieved. Therefore, a higher DTCE value is desired for an effective method.

Figure 8.8 shows a DTCE comparison for the methods shown in Table 8.2 as well as for the fan-out-aware method with fan-out count 3 and 4. From Figure 8.8, we can see that the proposed fan-out-aware method outperformed TA-SDD as well as previously published approaches. The DTCE value for the fan-out-aware method increased with an increase in fan-out count. For fairness of evaluation, Figure 8.9 shows the comparison between the proposed approach and TA-SDD for the same number of patterns.

In this case, the TA-SDD pattern set was truncated to match the pattern count obtained from the proposed method. Note that we based our comparison here on the proposed approach with a fan-out count greater than or equal to 2; however, similar results were obtained for higher fan-out counts as well. From Figure 8.9, we can see that, for all circuits, the proposed method always resulted in higher (maximum possible) TC. The truncated TA-SDD pattern set resulted in 0.01% to 0.85% loss in TC. Also, for a number of circuits, including the largest circuit (ethernet), the truncated TA-SDD pattern set resulted in lower DTC. For other circuits in which the truncated TA-SDD pattern set achieved a higher DTC, the improvement in DTC was slightly higher than the loss in TC.

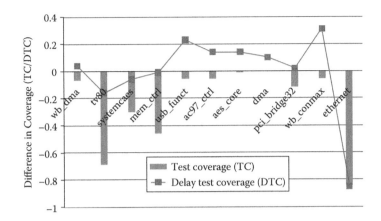

FIGURE 8.9
Difference in TC and DTC for truncated TA-SDD pattern set and proposed approach (fan-out ≥ 2).

8.4.2 Number of Unique Long Paths

The second metric we used to show the effectiveness of our approach was the number of long paths excited by the pattern set. This is an important metric because the excitation of a greater number of long paths implies possible detection of more SDDs. In the rest of the experiments, we only compared our approach with the default timing-aware pattern set (TA-SDD), which provided the maximum DTC value, as shown in the previous section.

For the five large IWLS circuits, Figure 8.10 shows the total number of unique paths with lengths greater than or equal to 90% of the clock period that were excited by the TA-SDD pattern set and the proposed method with different fan-out values. It can be noted from Figure 8.10 that the number of unique long paths excited by the fan-out-aware method decreased with an increase in fan-out count.

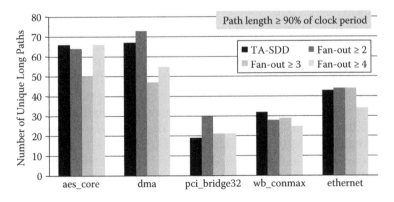

FIGURE 8.10
Comparison of the number of unique long paths between TA-SDD and fan-out-aware methods.

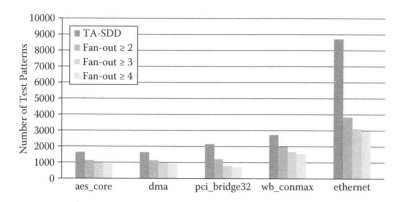

FIGURE 8.11
Comparison of test pattern count between TA-SDD patterns and the proposed method.

This is to be expected because the number of test patterns decreased with an increase in fan-out count, resulting in fewer excited paths. We can also see that the fan-out-aware method excited a larger number of long paths than the full timing-aware pattern set (TA-SDD). This is especially good considering the difference in the total number of patterns between TA-SDD and the fan-out-aware method (as shown in Figure 8.11). This also means the lower DTC value for the proposed method, as shown in Table 8.2, may have been caused by less coverage of short paths instead of long paths. To support this conclusion, Figure 8.12 shows the total number of unique paths with lengths greater than or equal to 10% of the clock period. From the figure, we can see that the proposed fan-out-aware method excited fewer such paths compared to TA-SDD, which confirms the conclusion that we derived previously.

For fair comparison, we also looked at the number of unique long paths detected by the TA-SDD patterns for the same pattern count as obtained by the fan-out-aware method. In this case, we truncated the TA-SDD pattern set

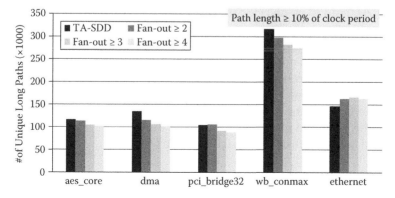

FIGURE 8.12
Comparison of number of unique short and long paths for TA-SDD and fan-out-aware methods.

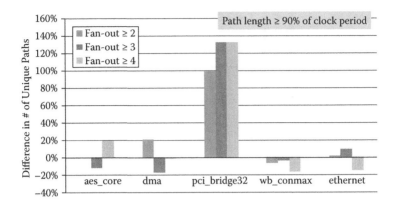

FIGURE 8.13
Comparison of the number of long paths between truncated TA-SDD and the proposed method.

to match the pattern count obtained from the fan-out method and calculated the number of excited unique long paths. Figure 8.13 shows the percentage difference in the number of unique long paths (length ≥ 90% of the clock period) obtained from the truncated TA-SDD pattern set and the corresponding fan-out-aware method.

From Figure 8.13, we can see that the fan-out-aware method outperformed the truncated TA-SDD pattern set. This is especially important from a volume production point of view since test pattern truncation may be required to fit the available vector memory on the automatic test equipment (ATE).

8.4.3 Length of Longest Path

The third metric we used to show the effectiveness of our approach was the length of the longest path excited by the pattern set. This is also an important metric because the excitation of the longest path is directly related to the size of the SDD that can be detected in a circuit given the system speed.

Figure 8.14 shows the comparison of the longest-path length excited by the TA-SDD pattern set and the proposed fan-out-aware method. Again, we can see that the proposed method excited longer or similar paths. Considering the difference in the overall pattern count, as shown in Figure 8.11, this is significant. Figure 8.15 shows the same comparison but for the truncated TA-SDD pattern set that had the same pattern count as the proposed method. We can see that the proposed method excited paths that were of similar length to or longer (2% to 6%) paths than the truncated TA-SDD set.

8.4.4 Number of Unique SDDs

Next, we compared the number of unique SDDs (gate pins) on the paths excited by the fan-out-aware and the TA-SDD patterns. We counted the

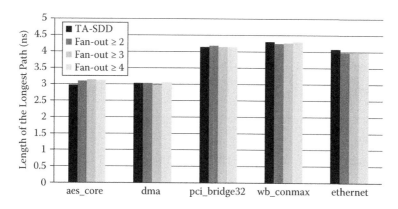

FIGURE 8.14
Longest-path length comparison for TA-SDD and the proposed method.

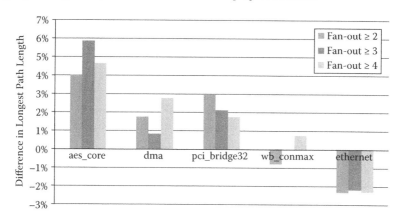

FIGURE 8.15
Longest sensitized path length comparison for truncated TA-SDD and the proposed method.

number of unique gate pins that were on the long paths excited by the pattern set. This represented the number of unique SDD locations that were on the long paths and covered by the pattern set. Figure 8.16a shows the number of unique SDDs (on paths with lengths greater than or equal to 90% of the clock period) for TA-SDD and the fan-out-aware method. From the figure, we can see that the fan-out-aware method resulted in a similar or higher number of unique SDDs. Compared to the truncated TA-SDD pattern set also, the proposed method resulted in a similar or higher unique number of SDDs, as shown in Figure 8.16b.

Based on the comparison using four different metrics, it can be concluded that the proposed fan-out-aware approach delivered significant reduction in pattern count and provided better or similar test quality for SDDs compared to the full or truncated timing-aware pattern set. Also, the proposed method

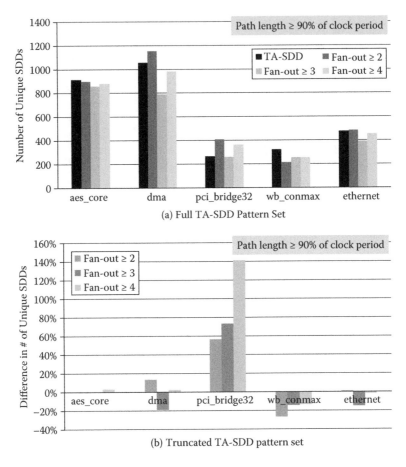

(a) Full TA-SDD Pattern Set

(b) Truncated TA-SDD pattern set

FIGURE 8.16
Comparison of number of unique SDDs between TA-SDD patterns and the proposed method.

achieved maximum coverage for the traditional transition fault coverage, which is not possible with the truncated TA-SDD pattern set. Another major advantage of the proposed method is very low computation time. Because only a few nodes are considered for time-consuming timing-aware pattern generation, the proposed method requires significantly less computation time (60% to 80%) than the TA-SDD pattern set, as shown in Figure 8.17.

8.4.5 Random Fault Injection and Detection

To show the added value of the SDD patterns over the traditional TDF patterns, as well as to show quality effectiveness of the proposed approach and TA-SDD patterns, we used random fault injection and verification. In our experiments, we injected a large number (50,000) of random SDDs of different sizes (defect size less than or equal to 10%–15% of clock period)

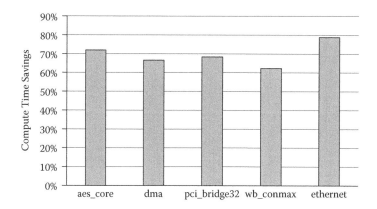

FIGURE 8.17
Computed time savings from the proposed method (fan-out ≥ 2) compared to TA-SDD.

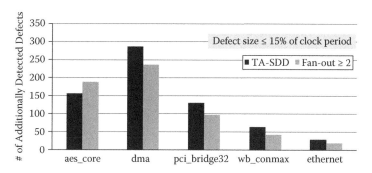

FIGURE 8.18
Number of additional detected SDDs for defect sizes 15% or less than the clock period.

on randomly selected nodes in each circuit. Note that we only inserted one defect at a time in a circuit. Next, fault simulation of TDF, full timing-aware ATPG patterns (TA-SDD), and the proposed fan-out-aware patterns for different fan-out values were carried out on each defective circuit.

For defect sizes less than or equal to 15% of the clock period, Figure 8.18 shows the number of defects that were detected in addition to the ones already detected by the TDF patterns. From the figure, we can see that not all injected defects could be detected by any of the pattern sets. This was to be expected since the defects were introduced randomly and therefore could also occur in the part of the circuit that was not excited by the pattern set. Also, SDDs injected on short paths were missed by most of the methods. From the figure, we can see that TA-SDD detected 30 to 286 additional defects, which means a 600- to 5,720-DPPM impact on quality. The fan-out-aware method, which has a significantly lower pattern count, detected 20 to 236 additional defects, showing a 400- to 4,720-DPPM impact. Also note that, for circuit aes_core, the fan-out-aware method detected more defects than TA-SDD.

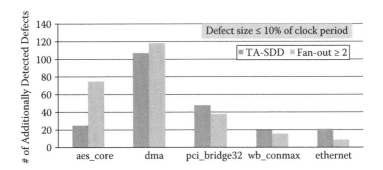

FIGURE 8.19
Number of additional detected SDDs for defect sizes 10% or less than the clock period.

For smaller defects, with defect size less than or equal to 10% of the clock period, the numbers of defects detected in addition for TA-SDD and fan-out-aware patterns are shown in Figure 8.19. In this case, we can clearly see the fan-out-aware method outperformed the TA-SDD pattern set. The quality impact of the fan-out-aware method was 180 to 2,360 DPPM, while for the full timing-aware SDD pattern set it was 420 to 2,160 DPPM.

To further understand the added quality value of SDD patterns, overlap between the different defects detected by TDF, TA-SDD, and the fan-out-aware method needs to be studied. Figure 8.20 shows the Venn diagrams for detected defects for defect sizes less than or equal to 15% of the clock period. Figure 8.21 shows the same but for defect sizes less than or equal to 10% of the clock period. From Figures 8.20 and 8.21, we can see that defect overlap between the TDF and the fan-out-aware patterns was larger than the overlap between the TDF and the SDD patterns. This was to be expected since the fan-out-aware pattern set contained a large number of TDF patterns and only a small number of timing-aware patterns. Also, note that the number of

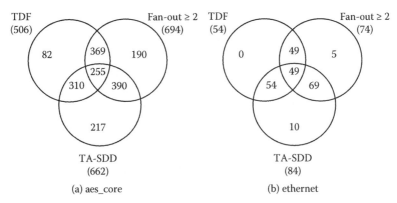

FIGURE 8.20
Detected defects overlap for defect sizes less than or equal to 15% of the clock period.

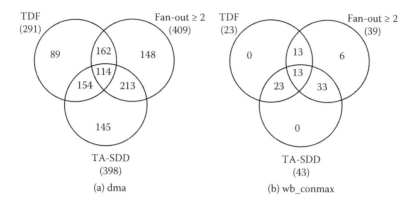

FIGURE 8.21
Detected defects overlap for defect sizes less than or equal to 10% of clock period.

defects uniquely detected only by the fan-out-aware method was similar to the number detected by full timing-aware patterns.

Because both TA-SDD and fan-out-aware methods resulted in a large number of unique defects, effective patterns can be selected from the two pattern sets to create the pattern set that has the highest quality. For effective pattern selection, the output deviation technique [Yilmaz 2008c] can be used.

8.5 Conclusion

Testing of SDDs is becoming a serious concern today, and the problem is likely to be aggravated for newer technologies. Existing solutions for testing SDDs are not practical enough for high-volume production environments due to large pattern count, large computation time, or both. In this chapter, we proposed a production-friendly method that takes the circuit topology into account while generating patterns for SDDs. The proposed method selects a small number of nodes in the circuit for which the timing-aware pattern generation needs to be carried out; for other nodes in the circuit, conventional transition fault patterns can be generated.

Experimental results on several IWLS'05 benchmark and six industrial circuits showed that, compared to the default timing-aware pattern set, the proposed method reduced the pattern count an average of 172% for IWLS benchmarks and an average of 105% for industrial circuits. The average penalty in absolute DTC was less than 1% for both IWLS and industrial circuits. However, by using other metrics, such as number of long paths, length of the longest path, and the number of unique SDDs, we have shown that the proposed method outperformed the default timing-aware ATPG, and the loss in

DTC was attributed to low coverage of short paths that are only important from a reliability point of view. Furthermore, random fault injection and verification of 50,000 SDDs showed that the proposed method achieved similar or higher quality for SDDs as default timing-aware ATPG and required only a small number of test patterns. This is important from the point of view of high-volume production.

References

[Ahmed 2006b] N. Ahmed, M. Tehranipoor, and V. Jayram, Timing-based delay test for screening small delay defects, in *Proceedings IEEE Design Automation Conference*, 2006.

[Devta-Prasanna 2009] N. Devta-Prasanna, S.K. Goel, A. Gunda, M. Ward, and P. Krishnamurthy, Accurate measurement of small delay defect coverage of test patterns, in *Proceedings International Test Conference*, 2009.

[Goel 2009] S.K. Goel, N. Devta-Prasanna, and R. Turakhia, Effective and efficient test pattern generation for small delay defects, in *Proceedings IEEE VLSI Test Symposium*, 2009.

[Goel 2010] S.K. Goel, K. Chakrabarty, M. Yilmaz, K. Peng, and M. Tehranipoor, Circuit topology-based test pattern generation for small delay defects, in *Proceedings IEEE Asian Test Symposium*, 2010.

[IWLS 2005] International Workshop on Logic and Synthesis, IWLS Benchmarks, 2005, http://iwls.org/iwls2005/benchmarks.html

[Kapur 2007] R. Kapur, J. Zejda, and T.W. Williams, Fundamental of timing information for test: How simple can we get?, in *Proceedings IEEE International Test Conference*, 2007.

[Kruseman 2004] B. Kruseman, A.K. Majhi, G. Gronthoud, and S. Eichenberger, On hazard-free patterns for fine-delay fault testing, in *Proceedings IEEE International Test Conference*, 2004.

[Li 2000] H. Li, Z. Li, and Y. Min, Reduction in number of path to be tested in delay testing, *Journal of Electronic Testing: Theory and Application [JETTA]* 16(5), pp. 477–485, 2000.

[Lin 2006] X. Lin, K.-H. Tsai, C. Wang, M. Kassab, J. Rajski, T. Kobayashi, R. Klingenberg, Y. Sato, S. Hamada, and T. Aikyo, Timing-aware ATPG for high quality at-speed testing of small delay defects, in *Proceedings IEEE Asian Test Symposium*, 2006, pp. 139–146.

[Lin 2007a] X. Lin, M. Kassab, and J. Rajski, Test generation for timing-critical transition faults, in *Proceedings IEEEE Asian Test Symposium*, 2007, pp. 493–500.

[Mitra 2004] S. Mitra, E. Volkerink, E.J. McCluskey, and S. Eichenberger, Delay defect screening using process monitor structures, in *Proceedings IEEE VLSI Test Symposium*, 2004.

[Pomeranz 1999] I. Pomeranz and S.M. Reddy, On n-detection test sets and variable n-detection test sets for transition faults, in *Proceedings IEEE VLSI Test Symposium*, 1999, pp. 173–180.

[Putman 2006] R. Putman and R. Gawde, Enhanced timing-based transition delay testing for small delay defects, in *Proceedings IEEE VLSI Test Symposium*, 2006.

[Sato 2005b] Y. Sato, S. Hamada, T. Maeda, A. Takatori, and S. Kajihara, Invisible delay quality—SDQM model lights up what could not be seen, in *Proceedings International Test Conference*, 2005, pp. 1202–1210.

[Turakhia 2007] R. Turakhia, W.R. Daasch, M. Ward, and J. van Slyke, Silicon evaluation of longest path avoidance testing for small delay defects, in *Proceedings IEEE International Test Conference*, 2007.

[Uzzaman 2006] A. Uzzaman, M. Tegethoff, B. Li, K. McCauley, S. Hamada, and Y. Sato, Not all delay tests are the same—SDQL model shows true-time, in *Proceedings IEEE Asian Test Symposium*, 2006, pp. 147–152.

[Waicukauski 1987] J.A. Waicukauski, E. Lindbloom, B.K. Rosen, and V.S. Iyengar, Transition fault simulation, *IEEE Design and Test of Computer*, 4(2), pp. 32–38, April 1987.

[Yan 2004] H. Yan and A.D. Singh, Evaluating the effectiveness of detecting delay defects in the slack interval: A simulation study, in *Proceedings IEEE International Test Conference*, 2004.

[Yilmaz 2008c] M. Yilmaz, K. Chakrabarty, and M. Tehranipoor, Test-pattern grading and pattern selection for small-delay defects, in *Proceedings IEEE VLSI Test Symposium*, 2008, pp. 233–239.

Section IV

SDD Metrics

9

Small-Delay Defect Coverage Metrics

Narendra Devta-Prasanna and Sandeep K. Goel

CONTENTS

9.1 Role of Coverage Metrics

Integrated circuits that are manufactured using advanced processes and designed to perform at high frequencies can suffer from subtle additional delays due to process variations and defects [Mitra 2004; Kim 2003; Kruseman 2004]. These delays are referred to as small-delay defects (SDDs), and they can affect the correct operation of a circuit at its rated speed. Several recent studies [Kruseman 2004; Turakhia 2007; Chang 2008; Yan 2009] showed that conventional tests based on a transition delay fault (TDF) model do not effectively screen all such defects. The need to achieve very high product quality and push toward zero defective parts per million (DPPM) for critical applications has stimulated research in the area of SDD testing.

Most methods of SDD testing can be broadly classified into two groups. The first approach is based on enhanced automatic test pattern generation (ATPG) techniques that incorporate circuit-timing information and generate tests for transition faults [Lin 2006; Mentor 2008; Synopsys 2008; Kajihara 2007]. These methods are also referred to as timing-aware ATPG. Conventional transition fault ATPG [Waicukauski 1987] targets gross-delay defects as it assumes that a delay defect is large enough to be detected irrespective of the propagation delay of the test path (sum of fault excitation and observation path). Therefore, these tests are generated without considering circuit delays. However, when SDDs are considered, then only delay defects that are bigger than the slack of the transition test path can be detected. Hence, timing-aware ATPG techniques try to improve the coverage of SDDs by generating transition fault patterns along longer test paths to minimize test slacks. One of the drawbacks of timing-aware ATPG is that it results in a high test pattern count. Published results [Yilmaz 2008a; Mattiuzzo 2008] showed that the size of the timing-aware test set can be up to eight times the size of a conventional transition fault test set. Some authors [Mattiuzzo 2008; Goel 2009, 2010; Lin 2007a] proposed different methods of reducing the size of the timing-aware SDD test set. The main idea in these methods is to use timing-aware ATPG for a subset of transition faults instead of the entire fault set and hence alleviate the problem of high pattern count.

The second approach for SDD testing is based on faster-than-at-speed test application [Kruseman 2004; Turakhia 2007; Putman 2006; Park 1992]. In this approach, transition fault tests are applied to the circuit at a clock speed faster than the rated operating frequency of the circuit. Since the slack of a test path is equal to the difference between the test clock period and the propagation delay of the test path, faster-than-at-speed techniques achieve higher SDD coverage by reducing the test clock period and thereby reducing the slack of the test paths. The main problem with this approach is that it can result in unnecessary yield loss due to hazards being captured as well as due to excessive IR drop and power consumption due to faster clocking. Also, some of the detected SDDs can be redundant defects if they are smaller than the slack of the longest functional path through the fault site. Methods for avoiding hazard capture during faster-than-at-speed testing are discussed elsewhere [Kruseman 2004; Turakhia 2007].

The path delay fault model can also be used to target SDDs, but it requires enumerating all paths in a circuit. Since the number of paths is huge and increases exponentially with circuit size, this approach is not practical [Li 2000]. Some other methods of SDD testing have been discussed [Uzzaman 2006; Yilmaz 2008b, 2008c]. Some methods proposed [Uzzaman 2006; Yilmaz 2008b, 2008c] are pattern-filtering methods that identify effective patterns for SDD testing from a set of conventional transition fault patterns.

The main objective of this chapter is to propose an accurate quantitative metric for measuring the SDD coverage of any given pattern set. In general, coverage metrics serve as useful figures of merit and fulfill several engineering and business needs, such as the following:

- Measuring the number (percentage) of defects that are covered by a given test
- Comparing different test generation and test application methods for their effectiveness at detecting defects
- Keeping track of test quality across different designs and products
- Specification of test quality requirements between customers and vendors
- Estimating the quality of the shipped product

In the past, several test quality metrics for SDD have been proposed. The statistical delay fault coverage (SDFC) [Park 1992; Kapur 2007] metric is based on the probability distribution of random and process-induced delay defects. The SDD coverage of a test set is computed as the ratio of all SDDs detected by the test set and the SDDs that could have been detected if all transition faults were tested along their longest testable paths. The main drawback of this method is that it assumes that tests are always applied at the system clock frequency. Hence, it cannot be used to accurately measure SDD coverage of tests that are applied at a frequency different from the system clock frequency.

The delay test coverage (DTC) metric [Lin 2006] measures test quality for SDDs by assigning a weight to each transition fault based on the ratio of the propagation delay of the tested path and propagation delay of the longest testable path through the fault. Similar to the SDFC metric, it does not take into account the frequency of the test application. Hence, it cannot be used to accurately measure or compare SDD coverage of tests when their application frequencies are different. Also, as we demonstrate in the next section, the DTC metric incorrectly assumes that delay defects of all sizes are equally likely to occur. However, several published studies [Tendolkar 1985; Nigh 2000] showed that smaller-delay defects are more likely to occur than bigger-delay defects.

The statistical delay quality level (SDQL) metric [Uzzaman 2006; Sato 2005a, 2005b; Kajihara 2007] addresses most of the shortcomings of the DTC and SDFC metrics. It takes into account the probability distribution of delay defects and the frequency of test application. However, the main problem with the SDQL metric is that it is a DPPM quality metric that can be used to estimate the number of test escapes and not the SDD coverage of a given test set. Furthermore, since the overall test escape rate is a function of several other fault models and test screens, the SDQL value cannot be used to estimate the final product quality (DPPM value). Instead, it can only be used to compare the relative difference in SDD test quality between two test sets. Also, we demonstrate in other sections that the SDQL value cannot be used to compare SDD test coverage of different products or to specify SDD coverage requirements between vendors and customers.

The SDD test coverage metric proposed in this chapter overcomes all the identified shortcomings of previously proposed metrics and satisfies all mentioned requirements of coverage metrics. The proposed method can also be used to report the quality and reliability components of SDD test coverage. The quality component refers to coverage of SDDs that are functionally irredundant and if undetected will result in a test escape; the reliability component refers to SDDs covered by the test that are functionally redundant at the time of manufacture but may result in field returns due to early life failure if left undetected. Our results also show that the proposed metric is several times faster to compute than the SDQL metric.

The rest of the chapter is organized as follows: In Section 9.2, we discuss the DTC and SDQL metrics. Section 9.3 describes the philosophy and the implementation of the proposed SDD coverage metric. Section 9.4 covers experimental results with several ISCAS89 benchmark circuits and some industrial designs. Section 9.5 concludes this chapter.

9.2 Overview of Existing Metrics

This section briefly describes the DTC and SDQL metrics for measuring SDD coverage and discusses their shortcomings as accurate and practical metrics.

9.2.1 Delay Test Coverage Metric

The DTC metric [Lin 2006] estimates SDD coverage by measuring the efficiency of a test set to test transition faults along the longest testable path through each fault. The metric value is calculated by taking the average weighted sum of all transition faults, with the weight assigned to each fault determined by the ratio of the propagation delay of the longest tested path and the propagation delay of the longest testable path through the fault site.

For example, consider the transition fault at the output of gate 3 in Figure 9.1. For simplicity, assume that the delay of each gate is 1 ns, all interconnect delays are zero, and all paths in the circuit are testable. The fault can be tested along paths that start at either FF1 or FF2 and end at FF3, FF4, or FF5. It can be seen that the longest path through the fault is that from FF1 (or FF2) to FF5; it has a delay of 6 ns. Now, if the fault is tested along this longest path, then its DTC weight is 6/6 = 1. If it is tested along the FF1-to-FF3 path, then its DTC weight is 4/6 = 0.667. If a transition fault is not detected by a test, then it has a weight of zero.

The DTC coverage of a design is calculated as follows:

$$DTC = \left[\sum_{i=1}^{F_{DS}} \left(\frac{L_t(i)}{L_{max}(i)} \right) + F_{DI} \right] \times \left[\frac{1}{F} \right] \times 100 \qquad (9.1)$$

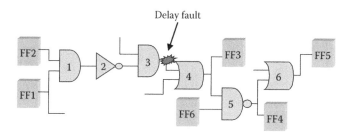

FIGURE 9.1
Example circuit for DTC metric.

where F_{DS} is the number of faults detected by the test, F_{DI} is the number of shift testable faults, and F is the total number of faults in the design. For each fault i, $L_{max}(i)$ is the delay of the longest testable path through fault i, while $L_t(i)$ is the delay of the longest tested path through fault i. Shift testable faults are associated with the scan operation of the circuit (e.g., scan enable and clock line faults). It can be deduced from above equation that maximum DTC coverage is achieved when all transition faults are tested along their longest testable paths. Also, the maximum DTC coverage value is equal to the TDF coverage of the test set.

9.2.1.1 Shortcomings of the DTC Metric

The DTC metric suffers from two major shortcomings. First, it does not consider the effect of test timing on SDD coverage. The DTC metric is based on the length (propagation delay) of test paths, whereas the number of delay defects detected by a test depends on the slack of the test path. Only delay defects larger than the test path slack can be detected by the test, and delay defects that are smaller than the test path slack remain undetected.

This is illustrated in Figure 9.2, where the transition fault at the output of gate 3 is tested along the path from FF1 to FF3 with a delay of 4 ns. If this test is applied to the circuit with a clock period of 7 ns, then all delay defects larger than 3 ns are detected. However, if the same test is applied with a clock period of 9 ns, then only delay defects larger than 5 ns are detected. Hence, the SDD coverage of the two tests is different, whereas the DTC value is the same since it only looks at test path lengths. Therefore, we can say that the DTC metric cannot be used to compare SDD coverage of two tests if they are applied at different frequencies.

The second problem with the DTC metric is that it is based on the assumption that delay defects of all sizes occur with equal probabilities. This is because, in DTC, a fault's weights are linearly proportional to their test path lengths. To understand this, consider the simple example of a transition fault that has a longest testable path length of 10 ns and a test clock period of 10 ns. If the fault is tested along the longest path, then all delay defects are tested,

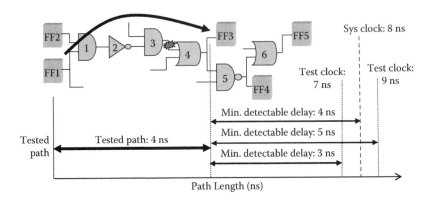

FIGURE 9.2
Effect of test timing on SDD coverage.

and the test has a DTC value of 100% for the fault. However, if the fault is tested along a 9-ns path, then all delay defects except those smaller than 1 ns are tested, and the test has a DTC value of 90%, implying that delay defects with a size between 0 and 1 ns constitute 10% of the defect population. Again, if the fault is detected along an 8-ns path, then the DTC value of the test is 80%, and all delay defects larger than 2 ns are detected. Therefore, it can be concluded that delay defects of size 1 to 2 ns are given the same weight as defects of size 0 to 1 ns. However, several published studies [Tendolkar 1985; Nigh 2000] based on silicon experiments showed that delay defects of different sizes have different probabilities, and in general, smaller-delay defects are much more likely to occur than larger-delay defects. The effect of this wrong assumption can be seen with the following example: Consider a simple circuit with four transition faults: f1, f2, f3, and f4. Assume that the test as well as the system clock period are 10 ns, and the longest testable path through each fault is also 10 ns. Also, assume that the probability distribution of delay defects is as shown in Figure 9.3.

This distribution is hypothetical, but it reflects the reality that smaller-delay defects are more likely than larger-delay defects. Now, consider two test sets that detect all four transition faults along different test paths as shown in Table 9.1.

The first column under each test set shows the length of the test path used for testing the fault, the second column shows the minimum size of the delay defect detected by the test based on the slack of the test path, and the third column shows the probability of a delay defect being detected at the fault site by the test based on the delay defect distribution of Figure 9.3. The last row shows the DTC value of the test set and the overall probability of a single delay defect being detected by the test set. It can be seen that even though test set 2 had higher DTC coverage, it had a lower probability of detecting a delay defect compared to test set 1. Therefore, the DTC metric does not

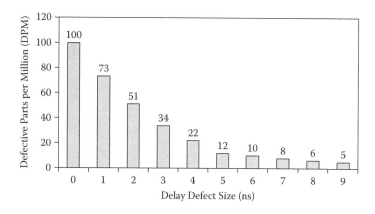

FIGURE 9.3
Delay defect distribution $H(s)$.

TABLE 9.1

DTC Value and Delay Defect Coverage

		Test Set 1			Test Set 2		
Fault	Longest Testable Path	Path Length	Minimum Defect Size	Defect Detection Probability	Path Length	Minimum Defect Size	Defect Detection Probability
f1	10	9	1	0.688	6	4	0.196
f2	10	2	8	0.034	4	6	0.090
f3	10	2	8	0.034	4	6	0.090
f4	10	2	8	0.034	4	6	0.090
		DTC = 37.5%			DTC = 45%		
		Average defect detection probability = 0.1975			Average defect detection probability = 0.1165		

accurately measure the SDD coverage of tests since it assumes uniform probability distribution for delay defects of all sizes.

9.2.2 Statistical Delay Quality Level Metric

The SDQL metric is based on the statistical delay quality model [Uzzaman 2006; Sato 2005a, 2005b; Kajihara 2007]. It is intended to reflect the outgoing delay quality of a product by measuring the number of delay defects that escape a test. The SDQL value of a fault is the undetected area under the curve between T_{mgn} and T_{det} (as shown in Figure 9.4), and it represents timing-irredundant defects that escape the test. In Figure 9.4, T_{det} represents the slack of the test path under the test clock period, and T_{mgn} represents the slack of the longest testable path through the fault under the system clock period. The area under the defect distribution curve to the right of T_{mgn} represents all delay defects

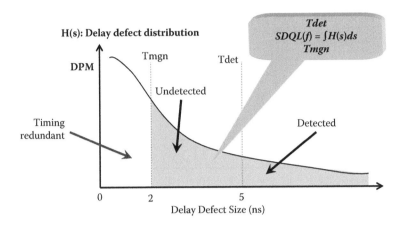

FIGURE 9.4
SDQL value of a fault.

that can result in a functional failure if left undetected, and the area to the right of T_{det} represents defects larger than the test path slack and hence the ones that are detected by the test.

For example, for the transition fault considered in Figure 9.2, T_{det} is 5 ns assuming a test clock period of 9 ns, and T_{mgn} is 2 ns assuming a system clock period of 8 ns. Therefore, the SDQL value of this fault for the defect distribution of Figure 9.3 is (51 + 34 + 22) = 107 DPPM. The SDQL value of an entire design is calculated by summing the SDQL value for each fault (both detected and undetected faults) in the design using the following equation:

$$\text{SDQL} = \sum_{i=1}^{F} \int_{T_{mgn}}^{T_{det}} H(s)\,ds \tag{9.2}$$

where F is the total number of faults in the design, and $H(s)$ is the delay defect distribution. It should be noted that since SDQL represents test escapes instead of defects covered by a test, a lower value of SDQL indicates higher SDD quality.

9.2.2.1 Shortcomings of the SDQL Metric

One of the drawbacks of the SDQL metric is that it is difficult to interpret. For example, if the SDQL value of a test is 215 DPM, it does not tell what percentage of SDDs are covered by the test. Whether an SDQL value of 215 DPM means detection of 90% of the SDDs or 60% of the SDD faults or x% of SDD faults, it is impossible to know. Also, the SDQL value is sensitive to the delay defect distribution, and as will be discussed further, even a small error in determining the exact shape and size of the distribution can sometimes result in a very big difference in the SDQL value.

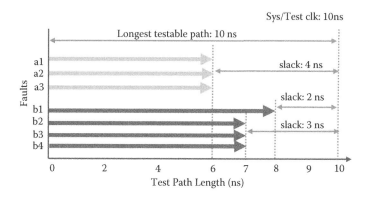

FIGURE 9.5
Transition fault test for designs A and B.

Another problem with the SDQL metric is that it is not normalized to the number of faults in a design. A bigger design will tend to have a higher (and hence inferior) SDQL value, which is understandable from a product quality point of view since bigger chips are more susceptible to failures. However, from a test coverage point of view, this can result in problems. When comparing the SDQL value of two different products, a lower SDQL value for one product does not always imply higher SDD coverage by its test even if the defect distribution, the system clock speed, and the test clock speed for the two products are the same.

For example, consider two designs: design A with three transition faults (a1 to a3) and design B with four transition faults (b1 to b4). Assume that the system as well as test clock period for both designs is 10 ns. Also, assume that both designs have the same delay defect distribution as shown in Figure 9.3. Assume that all the faults are detected by their respective tests along paths with lengths as shown in Figure 9.5. From the figure, we can see that T_{mgn} for all faults in both designs is 0 ns. T_{det} for all faults in design A is 4 ns, whereas the T_{det} value of fault b1 in design B is 2 ns; for the rest of the faults, it is 3 ns. The SDQL value for both designs can be calculated as follows:

$$\text{SDQL}(A) = \text{SDQL}(a_1) + \text{SDQL}(a_2) + \text{SDQL}(a_3)$$

$$= 3 \times \int_{0\,\text{ns}}^{4\,\text{ns}} H(s)\,ds = 3 \times (100 + 73 + 51 + 34) = 774\ \text{DPM}$$

$$\text{SDQL}(B) = \text{SDQL}(b_1) + \text{SDQL}(b_2) + \text{SDQL}(b_3) + \text{SDQL}(b_4)$$

$$= \int_{0\,\text{ns}}^{2\,\text{ns}} H(s)\,ds\ +\ 3 \times \int_{0\,\text{ns}}^{3\,\text{ns}} H(s)\,ds$$

$$= (100 + 73)\ +\ 3 \times (100 + 73 + 51) = 845\ \text{DPM}$$

We can see that the SDQL value of design A is better than the SDQL value of design B even though the design B test detects all faults along longer paths than the design A test; hence, the SDD coverage of the design B test is higher than the SDD coverage of the design A test. Therefore, we can conclude that the SDQL metric cannot be used to compare the SDD test coverage of two designs with a different number of faults.

Next, we discuss what happens when the SDQL metric is used to compare the SDD test coverage of designs with different operating frequencies. Since a design with more slack has more timing margin to tolerate SDDs, the resulting test escape rate and SDQL value for such designs tend to be lower. But, in the context of test coverage, poor SDD coverage of a design can be masked by a higher timing margin and result in a better SDQL value.

For example, consider two designs, each with three transition faults. Assume that the rated clock periods for design X and design Y are 10 ns and 12 ns, respectively. Also, assume that test patterns are applied to both designs at their rated frequencies. The longest testable path through all faults is 10 ns, and the faults are tested along paths with lengths as shown in Figure 9.6. It can be seen from the figure that the test for design X detects all faults along paths with smaller test slacks than the test path slacks for design Y faults. More precisely, the test for design X has a higher probability of catching an SDD than the design Y test.

This can be verified for the delay defect distribution of Figure 9.3. The probability that an SDD is detected by the design X test is the probability that a delay defect is 1 ns or bigger, which is equal to 0.688. Similarly, the probability that an SDD is detected by the design Y test is the probability that a delay defect is 4 ns or bigger while not considering defects smaller than 2 ns since such defects are timing redundant for this design. This probability is equal

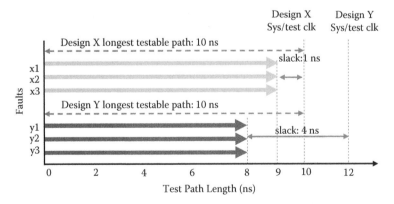

FIGURE 9.6
Transition fault test for modules X and Y.

to 0.426. Therefore, the design X test is superior to the design Y test in terms of catching SDDs. The SDQL for both designs can be calculated as follows:

$$SDQL(X) = SDQL(x_1) + SDQL(x_2) + SDQL(x_3)$$

$$= 3 \times \int_{0\,ns}^{1\,ns} H(s)\,ds = 3 \times 100 = 300 \text{ DPM}$$

$$SDQL(Y) = SDQL(y_1) + SDQL(y_2) + SDQL(y_3)$$

$$= 3 \times \int_{2\,ns}^{4\,ns} H(s)\,ds = 3 \times (51 + 34) = 255 \text{ DPM}$$

Here, it can be seen that design Y has a better SDQL value than design X even though the design X test has better SDD coverage than the design Y test. This is because SDQL takes into consideration the design-timing margin, which can mask poor SDD test coverage. Therefore, the SDQL metric cannot be reliably used for comparing SDD test coverage of different designs.

9.3 Proposed SDD Test Coverage Metric

In this section, we describe our proposed metric, referred to as the small-delay defect coverage (SDDC) metric [Devta-Prasanna 2009]. The SDDC value of a single fault is defined as the ratio of the probability that a delay defect at the fault site will be detected by the test and the probability that a delay defect is timing irredundant. In terms of path slacks and delay defect sizes, the SDDC value of a fault can be understood as the ratio of probability that a delay defect is larger than T_{det} and the probability that a delay defect is larger than T_{mgn}, where T_{det} is the slack of the longest tested path for the fault under the test clock, and T_{mgn} is the slack of the longest testable path through the fault site under the system clock. In terms of a delay defect distribution such as the one shown in Figure 9.4, the SDDC value of a fault can be viewed as the ratio of the "detected" area and the sum of the detected and "undetected" areas. Please note that the SDQL value of a fault is just the undetected area.

To compute the probability of a delay defect being bigger than a certain size, we transform the delay defect distribution $H(s)$ into the delay defect detection probability $G(s)$ in a preprocessing step. This transformation avoids the necessity to perform integration over $H(s)$ for every fault in the design, which is required for calculating the SDQL value. This step results

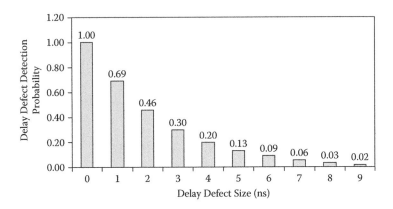

FIGURE 9.7
Delay defect detection probability $G(s)$.

in significant computation time savings, as will be shown by our results. The delay defect detection probability function $G(s)$ is obtained as follows:

$$G(s) = \frac{\int_s^\infty H(s)\,ds}{\int_0^\infty H(s)\,ds} \qquad \forall s \geq 0. \tag{9.3}$$

It should also be noted that $G(0) = 1$ and $G(s) = 0$ for all negative values of s. Conceptually, the difference between $H(s)$ and $G(s)$ functions can be stated as follows: $H(s)$ represents the probability that a delay defect of size s will occur; $G(s)$ represents the probability that the defect is of size s or more if a delay defect is already present. Figure 9.7 shows the $G(s)$ function for the $H(s)$ distribution shown in Figure 9.3.

The SDDC value of a design with a given test is calculated using Equation 9.4. The proposed metric considers all factors that affect SDD test coverage, such as test timing, test slack of fault detection path, and delay defect distribution; it accurately measures the percentage of timing-irredundant SDDs detected by a test. For at-speed tests, the maximum SDDC value is the same as the TDF coverage of a test, and it is achieved when all the faults are tested along their longest testable paths.

$$\text{SDDC} = \left[\left(\sum_{i=1}^{F_{DS}} \frac{G(T_{test} - l^i_{ptested})}{G(T_{system} - l^i_{pmax})} \right) + F_{DI} \right] \times \left[\frac{1}{F} \right] \times 100 \tag{9.4}$$

where $l^i_{ptested}$ is the length (propagation delay) of the longest tested path for fault i; l^i_{ipmax} is the length of the longest testable path through fault i; T_{system} is

TABLE 9.2

Comparison of SDQL and SDDC Values

Design	Number of Faults	System/Test Clock Period (ns)	SDQL (dpm)	SDDC (%)
A	3	10	774	20
B	4	10	845	34
X	3	10	300	69
Y	3	12	255	43.5

the system clock period; T_{test} is the test clock period; $G(s)$ is the delay defect detection probability for delay defect size s; F_{DS} is the number of faults detected by ATPG test patterns; F_{DI} is the number of shift testable faults; and F is the total number of faults in the chip.

Next, we compare SDDC and SDQL values in Table 9.2 for design A and B tests and design X and Y tests shown in Figure 9.5 and Figure 9.6, respectively, and discussed in the previous section. The $G(s)$ function shown in Figure 9.7 is used to compute the SDDC values. It can be seen from column 5 in the table that design B had a higher SDDC value (34%) than design A (20%), which confirms our analysis in the previous section that the design B test had a higher coverage of SDDs. Similarly, design X had a higher SDDC value (69%) than design Y (43.5%), which also confirms our analysis in the previous section.

These examples demonstrate that the SDDC metric accurately measured SDD test coverage and can be used to compare SDD coverage of designs with a different number of faults or different operating speeds. In the experimental results section, we show more results that compare SDDC and SDQL values for several benchmark circuits. We also give analysis of the effect of system speed (timing margin) on SDQL and SDDC metrics.

9.3.1 Quadratic SDD Test Coverage Metric

In this section, we propose a method for estimating SDD test coverage when defect distribution data are not available. The proposed method is based on the length of test paths similar to the DTC metric, but it also takes into account the effect of test timing on SDD coverage and gives greater credit to the detection of smaller-delay defects than larger-delay defects. To distinguish this method from the SDDC metric proposed in the previous section, we refer to it as quadratic SDDC ($SDDC^Q$) metric since the weight assigned to each fault is proportional to the square of its longest tested path length. The $SDDC^Q$ value of a test is calculated using the following equation:

$$SDDC^Q = \left[\left(\sum_{i=1}^{F_{DS}} \frac{(P_i)^2}{(P_{maxi})^2}\right) + F_{DI}\right] \times \left[\frac{1}{F}\right] \times 100. \qquad (9.5)$$

For example, the SDDC^Q values of design A and design B tests shown in Figure 9.5 are 36% and 52.75%, respectively; for design X and design Y tests shown in Figure 9.6, their SDDC^Q metric values are 81% and 64%, respectively. Please note that other variations of this method are also possible, with weights assigned to faults based on the cube or some other function of the test path lengths.

9.3.2 Faster-than-at-Speed Testing

In this section, we discuss the use of the SDDC metric to evaluate SDD coverage of faster-than-at-speed tests. In general, when a test is applied at a faster rate than the system speed, the slack of test paths is reduced; hence, SDD coverage of the test is improved. Here, we discuss the handling of two special cases.

1. Masking of faults: It is possible that due to faster clocking, the length of the longest tested path for some faults may exceed the test clock period. Such paths have negative slacks, and the corresponding observation flops should be masked to avoid false failures. While calculating the SDDC metric, the SDDC value of such faults becomes zero since $G(s)$ for negative slack is zero. However, if some of these faults are also tested by other patterns in the test along paths that do not exceed the test clock period, the SDD coverage contribution of those faults need to be considered. As the timing report generated by the ATPG tool used in our experiments only contained the longest tested path for each fault, SDDC coverage values reported in this chapter for faster-than-at-speed tests is pessimistic. Ideally, SDDC computation should be done by the ATPG tool, and when a faster-than-at-speed clock rate is specified, the ATPG tool should consider the SDD coverage of the longest test path for each fault that can be observed for the specified clock period.

2. Detection of timing-redundant SDDs: Timing-redundant SDDs are tested when the T_{det} value of a fault is less than its T_{mgn} value due to a faster-than-at-speed test clock. For example, T_{mgn} for design Y faults shown in Figure 9.6 is 2 ns, assuming the system clock period is 12 ns. Therefore, all delay defects smaller than 2 ns are timing redundant since their presence does not result in any functional failure. However, if the test shown in the figure is applied to the circuit at 9 ns, then the test will detect any delay defect that is larger than 1 ns, including timing-redundant delay defects in the range of 1 to 2 ns. Detection of such defects is commonly referred to as reliability testing since even though they do not result in time zero failures, they pose a reliability risk if the defect deteriorates over time and causes an early life failure. With the proposed SDDC metric, it is possible to quantitatively measure the coverage of reliability delay defects. For

example, if the SDDC value of a fault is $G(1\ ns)/G(2\ ns) = 0.69/0.46 = 1.5$, then this is the total SDDC value of the fault, and it consists of both DPPM and reliability components of SDD coverage. Since the maximum value of the DPPM component in the SDDC value of a fault is 1 $[G(T_{det} = T_{mgn})/G(T_{mgn}) = 1]$, the remainder of the SDDC value (i.e., $1.5 - 1 = 0.5$) corresponds to the reliability SDD component.

When the SDDC metric is computed, the number of reliability delay defects detected by a test is recorded and expressed as a percentage of the total timing-irredundant delay defects in the circuit. This value is referred to as $SDDC_{EFR}$ coverage, and the coverage of remaining timing-irredundant delay defects detected by the test is referred to as $SDDC_{DPM}$ coverage. For example, if the test for design Y shown in Figure 9.6 is applied at a clock rate of 9 ns (faster than at speed), then the total SDDC coverage is 150%. This means $SDDC_{DPM}$ coverage is 100%; the $SDDC_{EFR}$ is 50%. It should be noted that for at-speed and slower-than-at-speed tests, $SDDC_{EFR}$ coverage is always zero, and $SDDC_{DPM}$ coverage is equal to the total SDDC coverage for the test.

9.4 Experimental Results

Here, we present experimental results for 12 benchmark circuits that include 8 full-scan versions of ISCAS89 circuits and 4 industrial designs (named A, B, C, and D). The timing information for the ISCAS89 circuits was generated such that path propagation delays were based on the number of gates in a path and the fan-out of gate outputs, and the system clock period was determined as 1.2 times the delay of the longest path in the circuit. Then, the timing was normalized to a system clock period of 10 ns. The system clock periods for circuits A, B, C, and D were 4.0, 4.7, 4.7, and 9.4 ns, respectively. For each circuit, we generated two sets of patterns using the FastScan commercial ATPG tool [Mentor 2008]. The first set was generated using conventional transition fault ATPG, and the second set was generated using timing-aware ATPG with a slack margin value of 5%. Slack margin value is an input parameter to the ATPG tool, and it is used to control the SDD quality of patterns [Lin 2006]. Table 9.3 shows some characteristics of the benchmark circuits.

The first column shows the circuit name. The second and third columns show the total number of faults and the number of shift testable faults, respectively. Columns 4 to 7 show the pattern count and TDF coverage for the two test sets. The last row shows the total number of patterns for all benchmark circuits.

Timing reports containing information on the length of the longest tested path and the longest testable path for each fault were generated by performing timing-aware fault simulation of the two test sets for each circuit.

TABLE 9.3

Characteristics of Benchmark Circuits

Circuit	F	F_{DI}	Conventional TDF		Timing-Aware TDF	
			Patterns	TDF (%)	Patterns	TDF (%)
s1423	4,434	600	109	95.17	207	95.17
s5378	15,948	1,436	203	92.28	262	92.29
s9234	29,518	1,809	387	86.24	706	86.24
s13207	45,254	5,311	340	84.64	590	84.64
s15850	53,034	4,682	214	76.56	502	76.56
s35932	106,690	13,860	76	90.52	313	90.52
s38417	125,066	13,112	241	98.92	636	98.93
s38584	119,154	11,624	386	89.43	699	89.43
A	66,261	11,440	341	88.27	707	88.29
B	83,591	14,312	161	90.44	454	90.44
C	647,019	66,043	2,018	89.21	16,724	89.21
D	15,341,45	176,480	6,950	83.13	32,384	83.13
Total			11,426		54,184 (4.7X)	

Then, we calculated the DTC, SDQL, SDDC, and SDDC^Q metric values using Equations 9.1, 9.2, 9.4, and 9.5, respectively. For delay defect distribution, we used

$$H(s) = 1.58 \times 10^{-3} \times e^{-2.1s} + 4.94 \times 10^{-6} \text{ (dpm)} \qquad (9.6)$$

which is based on the defect distribution data given by Mitra et al. [Mitra 2004] and subsequently used to calculate SDQL results [Lin 2006; Uzzaman 2006; Sato 2005b].

Table 9.4 shows the comparison between different SDD coverage metrics; it is organized as follows: The first column shows the circuit name. Columns 2 to 5 show the DTC, SDQL, SDDC, and SDDC^Q values, respectively, for the traditional TDF test; columns 6 to 9 show the same for the timing-aware TDF test. It can be seen from the table that the SDQL metric cannot be used to accurately compare the SDD test coverage of different designs. For example, the SDQL value of the traditional TDF test for circuit s1423 was 0.0563, which is almost twice better than the SDQL value of 0.1083 for s5378. However, s5378 had a higher SDD test coverage than s1423 since the s5378 test had an SDDC value of 90.10%, which is higher than the SDDC value of 84.15% for s1423.

Similarly, the DTC metric cannot be used to accurately compare the SDD coverage of two tests as it gives equal weight to delay defects of all sizes. For example, if we compare the DTC values of the traditional TDF tests for circuits s35932 and s38584, then s35932 had a 0.16% higher DTC coverage. But, based on the SDDC^Q metric that gives proportionally higher credit to the detection

TABLE 9.4

SDD Test Coverage Using Different Metrics

Circuit	Conventional TDF Test				Timing-Aware TDF Test (5% *sm*)			
	DTC (%)	SDQL	SDDC (%)	SDDC^Q (%)	DTC (%)	SDQL	SDDC (%)	SDDC^Q (%)
s1423	68.82	0.0563	84.15	58.46	74.17	0.0494	85.99	64.27
s5378	88.43	0.1083	90.10	85.68	89.23	0.1015	90.56	87.04
s9234	74.21	0.4173	80.62	66.80	76.21	0.3889	81.62	69.91
s13207	78.40	0.5494	82.06	73.94	78.83	0.5442	82.20	74.63
s15850	64.96	1.0146	71.80	58.34	66.08	0.9993	72.07	59.64
s35932	81.41	1.8064	82.76	74.73	83.29	1.7102	83.66	77.58
s38417	92.35	0.2915	96.68	87.49	94.53	0.2231	97.41	91.09
s38584	81.25	0.8434	86.69	76.07	82.15	0.8251	86.92	77.52
A	73.17	1.1158	81.05	63.22	74.21	1.105	81.25	64.78
B	76.62	5.3044	80.39	71.11	77.49	5.2926	80.55	72.48
C	75.64	16.623	76.65	65.47	81.52	13.574	80.13	74.60
D	64.48	57.24	70.98	53.00	70.13	55.9075	72.90	59.93

of smaller-delay defects, it can be seen that the SDD coverage of s35932 test was actually less than the s38584 test by 1.34%. In another example, if we compare the DTC coverage of traditional as well as the timing-aware TDF tests for circuit s9234, then the timing-aware test had 2% higher DTC coverage whereas if we look at the SDDC^Q metric, then the timing-aware test had a 3.11% higher SDD coverage than the traditional TDF test.

Table 9.5 shows the CPU (central processing unit) run time in seconds required for computing different SDD metrics. Faster CPUs were used to compute the metrics for industrial circuits to save time. It can be seen from this table that the DTC metric was the fastest to compute, the SDQL metric was the most computation intensive, and the run times for the SDDC and SDDC^Q metrics were comparable to the DTC metric. On average, the SDQL metric required 21.7 times more computation time than the DTC metric, whereas the SDDC and SDDC^Q metrics required 1.4 and 1.2 times more computation time than the DTC metric, respectively.

9.4.1 Sensitivity to System Frequency

In this section, we examine the sensitivity of the SDQL and SDDC metrics to the system clock period. As mentioned, coverage metrics are also used to specify test quality requirements between clients and vendors. For example, a client may ask the vendor to supply a product designed to operate at a certain frequency and that the product be tested to a certain level with respect to SDD quality. However, it might happen that the vendor cannot close design timing at the specified operating frequency but can come very close to the target.

TABLE 9.5

Run Time to Compute SDD Metrics

	Conventional TDF Test				Timing-Aware TDF Test (5% *sm*)			
Circuit	DTC	SDQL	SDDC	SDDC^Q	DTC	SDQL	SDDC	SDDC^Q
s1423	1	18	4	1	1	18	4	1
s5378	2	25	4	2	2	23	5	3
s9234	5	103	9	5	5	98	9	5
s13207	7	138	12	7	7	133	12	8
s15850	8	240	15	8	8	230	14	8
s35932	16	212	20	17	16	204	20	17
s38417	20	274	26	20	19	247	26	20
s38584	18	308	25	18	18	289	25	19
A	15	225	15	30	15	195	30	30
B	15	225	30	30	30	225	30	30
C	135	3,000	300	240	120	2,580	135	210
D	240	10,695	600	450	255	9,525	315	495
Total	482	15,463	1,060	828	496	13,767	625	846
Inc.	1X	21.7X	1.4X	1.2X	1X	20X	1.4X	1.3X

To examine the impact of using SDQL and SDDC metrics to specify test quality requirements in such situations, we selected three circuits (s1423, s9234,
and s38584) and evaluated the SDQL and SDDC values for their conventional
TDF test by assuming that the circuits were operated at different system frequencies. It was also assumed that for each value of the system frequency,
the test was applied at the same rate as system clock. Also, the same defect
distribution given in Equation 9.6 was used in all cases. Figure 9.8 shows the
percentage change in SDQL value for the three circuits for different values
of system/test clock periods when compared to a system/test clock period of

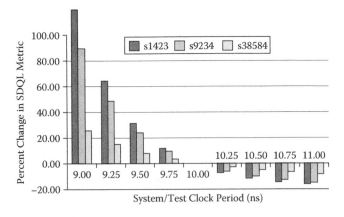

FIGURE 9.8
SDQL value for different system clock periods.

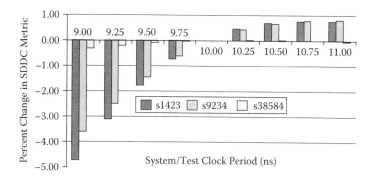

FIGURE 9.9
SDDC value for different system clock periods.

10 ns. Similarly, Figure 9.9 shows the percentage change in SDDC value of the circuits. It can be seen from Figure 9.8 that the SDQL value of a test can vary by as much as +120% to –15% for a ±10% change in system frequency and by +30% to –15% for a ±5% change in system frequency. On the other hand, the SDDC metric was relatively robust, and its value varied by less than 6% for a ±10% change in system frequency. In the case of s38584, the SDDC value of the test changed by less than 0.3% when the system clock period was varied by ±10%. Therefore, it can be concluded that the SDDC metric is a more reliable method of specifying SDD test coverage requirements.

9.4.2 Sensitivity to Defect Distribution

In this section, we examine the sensitivity of the SDQL and SDDC metrics to changes or inaccuracies in determining the exact distribution of delay defects. This is important to study since delay defect distributions are susceptible to changes over time; also, it is difficult to exactly determine the shape and size of the distribution. Figure 9.10 shows six different delay defect distributions whose mathematical functions are given next. The inside graphs in the figure show the distributions more closely in the 0- to 4-ns range.

$$H1(s) = 1.58 \times 10^{-3} \times e{-2.1s} + 4.94 \times 10^{-6}$$

$$H2(s) = 0.79 \times 10^{-3} \times e{-2.1s} + 2.47 \times 10^{-6}$$

$$H3(s) = 1.78 \times 10^{-3} \times e{-2.0s} + 10.0 \times 10^{-6}$$

$$H4(s) = 1.58 \times 10^{-3} \times e{-1.8s} + 4.94 \times 10^{-6}$$

$$H5(s) = 1.50 \times 10^{-3} \times e{-2.2s} + 10.0 \times 10^{-6}$$

$$H6(s) = 1.18 \times 10^{-3} \times e{-2.5s} + 20.0 \times 10^{-6}$$

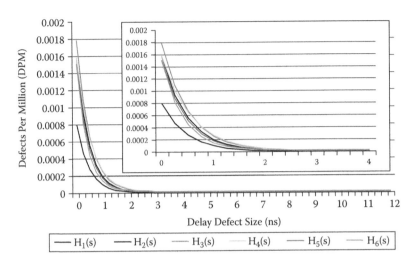

FIGURE 9.10
Delay defect distributions.

$H1(s)$ is the distribution in Equation 9.6 that we used for generating the results shown in Table 9.4. $H2(s)$ is obtained by dividing $H1(s)$ by two. Therefore, it has the same shape as $H1(s)$ and differs only in magnitude. $H3(s)$ to $H6(s)$ were obtained by making small changes in the parameters for $H1(s)$. Again, we selected three test case circuits and evaluated the SDQL and SDDC values of their conventional TDF test for each of the six defect distributions.

Figure 9.11 shows the percentage change in SDQL value of the test for the different distributions with respect to the SDQL value obtained for $H1(s)$. It can be seen that for $H2(s)$, the SDQL value was half the SDQL value of $H1(s)$.

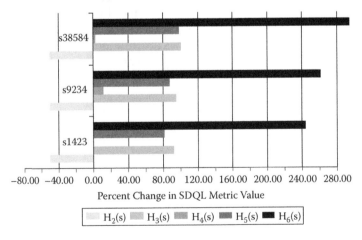

FIGURE 9.11
SDQL value for different defect distributions.

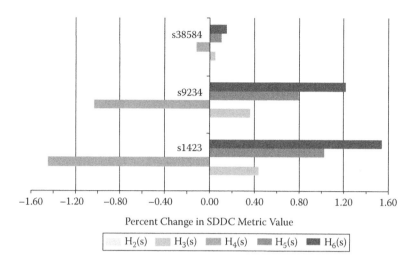

FIGURE 9.12
SDDC value for different defect distributions.

For other distributions, the SDQL value varied by up to 280%. Similarly, Figure 9.12 shows the percentage change in SDDC value of the tests. It can be seen from the figure that the SDDC value changed by less than ±2% for the different defect distributions, and in the case of s38584, it changed by less than ±0.2%. Also, the SDDC value for $H2(s)$ is exactly the same as the SDDC value for $H1(s)$ since both distributions have the same shape. Therefore, the SDDC metric was more robust in the face of small changes to the delay defect distribution, and it more accurately reflected the real SDD test coverage even if the delay defect distribution was not completely accurate.

9.4.3 Timing-Aware versus Faster-than-at-Speed

Here, we discuss the results of our experiments comparing two techniques of improving SDD test coverage: (1) timing-aware ATPG and (2) faster-than-at-speed test application. For each benchmark circuit, we evaluated the SDDC coverage of both conventional and timing-aware tests at five test application frequencies: 5% faster than the test clock, 2.5% faster than the test clock, at the speed of the test clock ($T_{test} = T_{sys}$), 2.5% slower than the test clock, and 5% slower than the test clock compared to the fixed system clock. In each case, we also report the $SDDC_{DPM}$ and $SDDC_{EFR}$ coverage of the tests and the corresponding SDQL values. Experimental results for six benchmark circuits are shown in Table 9.6.

In the table, the first column shows the circuit name, and the second column shows the percentage difference between test clock period and system clock period, $(T_{sys} - T_{test})/T_{sys}$. A positive difference means faster-than-at-speed testing, and a negative difference indicates slower-than-at-speed testing.

TABLE 9.6

SDD Coverage for Timing-Aware Tests and Faster-than-at-Speed Testing

Circuit	T_{test} (%)	Conventional TDF Test				Timing-Aware TDF Test (5% sm)			
		SDDC	SDDC$_{DPM}$	SDDC$_{EFR}$	SDQL	SDDC	SDDC$_{DPM}$	SDDC$_{EFR}$	SDQL
s35932	+5	86.06	84.73	1.33	1.3735	87.29	85.92	1.37	1.3772
	+2.5	84.31	83.67	0.64	1.557	85.33	84.66	0.67	1.5602
	0	82.76	82.76	0	1.8064	83.66	83.66	0	1.7102
	−2.5	81.35	81.35	0	1.8378	82.16	82.16	0	1.8407
	−5	80.00	80.00	0	2.0426	80.77	80.77	0	1.9596
s38417	+5	100.04	97.78	2.26	−0.056	100.79	98.30	2.49	−0.055
	+2.5	98.36	97.34	1.02	0.0834	99.09	97.97	1.12	0.0849
	0	96.68	96.68	0	0.2915	97.41	97.41	0	0.2231
	−2.5	95.02	95.02	0	0.359	95.74	95.74	0	0.3607
	−5	93.35	93.35	0	0.5657	94.07	94.07	0	0.4978
s38584	+5	90.06	87.84	2.22	0.6033	90.31	87.98	2.33	0.583
	+2.5	88.43	87.39	1.04	0.724	88.66	87.56	1.10	0.705
	0	86.69	86.69	0	0.8434	86.92	86.92	0	0.8251
	−2.5	85.08	85.08	0	0.9611	85.30	85.30	0	0.9433
	−5	83.51	83.51	0	1.0783	83.73	83.73	0	1.0608
B	+5	81.79	80.88	0.91	5.1905	81.99	80.99	1.00	5.1756
	+2.5	81.04	80.68	0.36	5.2505	81.22	80.82	0.40	5.2373
	0	80.39	80.39	0	5.3044	80.55	80.55	0	5.2926
	−2.5	79.76	79.76	0	5.3577	79.90	79.90	0	5.3469
	−5	79.17	79.17	0	5.4072	79.30	79.30	0	5.3973
C	+5	81.62	80.30	1.32	12.736	86.95	84.07	2.88	8.0016
	+2.5	79.26	78.73	0.53	14.816	83.59	82.56	1.03	11.016
	0	76.65	76.65	0	16.623	80.13	80.13	0	13.574
	−2.5	75.10	75.10	0	17.982	77.98	77.98	0	15.493
	−5	73.80	73.80	0	19.114	76.21	76.21	0	17.062
D	+5	74.04	72.88	1.16	54.261	76.18	74.76	1.42	52.592
	+2.5	72.89	72.31	0.58	55.714	74.98	74.26	0.72	54.213
	0	70.98	70.98	0	57.24	72.90	72.90	0	55.908
	−2.5	69.90	69.90	0	58.737	71.78	71.78	0	57.506
	−5	68.84	68.84	0	60.149	70.70	70.70	0	58.984

Subsequent columns show SDDC, SDDC$_{DPM}$, and SDDC$_{EFR}$ coverage as well as SDQL values for both the conventional TDF test and the timing-aware test. Based on these results, we can draw the following conclusions:

1. If only a limited number of patterns can be applied on the tester, then faster-than-at-speed testing offers the most cost-effective method for improving SDD coverage rather than applying timing-aware ATPG patterns. In general, timing-aware ATPG requires several times more patterns than conventional TDF ATPG. On average, it requires 4.7

times more patterns. However, in terms of SDD coverage improvement, just 5% overclocking of the conventional TDF patterns has higher $SDDC_{DPM}$ coverage than timing-aware tests. In addition, a small amount of $SDDC_{EFR}$ coverage is obtained. If only 2.5% overclocking is done, then $SDDC_{DPM}$ coverage of the test is still comparable and in some cases higher than the SDDC value of timing-aware tests. Please note that the same conclusion can be derived if SDQL values of the tests are compared.

2. If the goal is to achieve maximum SDD test coverage, then the best results are obtained by generating timing-aware test patterns and applying them faster than at speed.

3. It is important to accurately control test timing. If the test clock is slow by even 2.5%, then SDD coverage can be improved by a comparable amount by correctly clocking the test rather than generating timing-aware test patterns. Similarly, when timing-aware tests are applied, the additional SDD coverage gained by applying these tests can be lost if the test clock is slower by even 2.5%.

9.5 Conclusion

Small-delay defect testing may be necessary to maintain high quality for parts manufactured using advanced processes and designed to operate at high frequencies. In this chapter, we studied the subject of coverage metrics for measuring test coverage of SDDs. We critically examined previously proposed SDD coverage metrics and concluded that they do not fulfill all the requirements of practical and accurate test coverage metrics. We proposed new metrics for measuring SDD test coverage that overcome the shortcomings of previously proposed metrics. We explained the philosophy behind the new metrics and presented the results of our experiments for several academic and industrial benchmark circuits. We also compared the effectiveness of two SDD testing strategies, timing-aware ATPG and faster-than-at-speed testing, in improving the SDD coverage of tests.

References

[Ahmed 2006b] N. Ahmed, M. Tehranipoor, and V. Jayram, Timing-based delay test for screening small delay defects, in *Proceedings IEEE Design Automation Conference*, 2006.

[Chang 2008] C.-J. Chang and T. Kobayashi, Test quality improvement with timing-aware ATPG: Screening small delay defect case study, poster session, International Test Conference, 2008.

[Devta-Prasanna 2009] N. Devta-Prasanna, S.K. Goel, A. Gunda, M. Ward, and P. Krishnamurthy, Accurate measurement of small delay defect coverage of test patterns, in *Proceedings International Test Conference*, 2009.

[Goel 2009] S.K. Goel, N. Devta-Prasanna, and R. Turakhia, Effective and efficient test pattern generation for small delay defects, in *Proceedings IEEE VLSI Test Symposium*, 2009.

[Goel 2010] S.K. Goel, K. Chakrabarty, M. Yilmaz, K. Peng, and M. Tehranipoor, Circuit topology-based test pattern generation for small delay defects, in *Proceedings IEEE Asian Test Symposium*, 2010.

[Kajihara 2007] S. Kajihara, S. Morishima, M. Yamamoto, X. Wen, M. Fukunaga, K. Hatayama, and T. Aikyo, Estimation of delay test quality and its application to test generation, in *Proceedings IEEE International Conference on Computer-Aided Design*, 2007.

[Kapur 2007] R. Kapur, J. Zejda, and T.W. Williams, Fundamental of timing information for test: How simple can we get?, in *Proceedings IEEE International Test Conference*, 2007.

[Kim 2003] K.S. Kim, S. Mitra, and P.G. Ryan, Delay defect characteristics and testing strategies, in *IEEE Design and Test of Computers*, September–October 2003.

[Kruseman 2004] B. Kruseman, A.K. Majhi, G. Gronthoud, and S. Eichenberger, On hazard-free patterns for fine-delay fault testing, in *Proceedings IEEE International Test Conference*, 2004.

[Li 2000] H. Li, Z. Li, and Y. Min, Reduction in number of path to be tested in delay testing, *Journal of Electronic Testing: Theory and Application [JETTA]* 16(5), 477–485, 2000.

[Lin 2006] X. Lin, K.-H. Tsai, C. Wang, M. Kassab, J. Rajski, T. Kobayashi, R. Klingenberg, Y. Sato, S. Hamada, and T. Aikyo, Timing-aware ATPG for high quality at-speed testing of small delay defects, in *Proceedings IEEE Asian Test Symposium*, 2006, pp. 139–146.

[Lin 2007a] X. Lin, M. Kassab, and J. Rajski, Test generation for timing-critical transition faults, in *Proceedings IEEEE Asian Test Symposium*, 2007, pp. 493–500.

[Mattiuzzo 2008] R. Mattiuzzo D. Appello, and C. Allsup, Small delay defect testing, in *Proceedings Synopsys User's Group Conference*, 2008.

[Mentor 2008] *Scan and ATPG Process Guide, FastScan Tool*, Application Note, August 2008.

[Mitra 2004] S. Mitra, E. Volkerink, E.J. McCluskey, and S. Eichenberger, Delay defect screening using process monitor structures, in *Proceedings IEEE VLSI Test Symposium*, 2004.

[Nigh 2000] P. Nigh and A. Gattiker, Test method evaluation experiments and data, in *Proceedings IEEE International Test Conference*, 2000, pp. 454–463.

[Park 1992] E.S. Park, M.R. Mercer, and T.W. Williams, The total delay fault model and statistical delay fault coverage, in *IEEE Transactions on Computers* 41(6), pp. 688–698, June 1992.

[Putman 2006] R. Putman and R. Gawde, Enhanced timing-based transition delay testing for small delay defects, in *Proceedings IEEE VLSI Test Symposium*, 2006.

[Sato 2005a] Y. Sato, S. Hamada, T. Maeda, A. Takatori, and S. Kajihara, Evaluation of the statistical delay quality model, in *Proceedings of Asia and South Pacific Design Automation Conference*, 2005, pp. 305–310.

[Sato 2005b] Y. Sato, S. Hamada, T. Maeda, A. Takatori, and S. Kajihara, Invisible delay quality—SDQM model lights up what could not be seen, in *Proceedings International Test Conference*, 2005, pp. 1202–1210.

[Synopsys 2008] Synopsis Inc., *TetraMax ATPG User Guide*, Synopsis Inc., Mountain View, CA, 2008.

[Tendolkar 1985] N. Tendolkar, Analysis of timing failures due to random AC defects in VLSI modules, in *Proceedings IEEE Design Automation Conference*, 1985.

[Turakhia 2007] R. Turakhia, W.R. Daasch, M. Ward, and J. van Slyke, Silicon evaluation of longest path avoidance testing for small delay defects, in *Proceedings IEEE International Test Conference*, 2007.

[Uzzaman 2006] A. Uzzaman, M. Tegethoff, B. Li, K. McCauley, S. Hamada, and Y. Sato, Not all delay tests are the same—SDQL model shows true-time, in *Proceedings IEEE Asian Test Symposium*, 2006, pp. 147–152.

[Waicukauski 1987] J.A. Waicukauski, E. Lindbloom, B.K. Rosen, and V.S. Iyengar, Transition fault simulation, *IEEE Design and Test of Computer*, pp. 32–38, April 1987.

[Yan 2004] H. Yan and A.D. Singh, Evaluating the effectiveness of detecting delay defects in the slack interval: A simulation study, in *Proceedings IEEE International Test Conference*, 2004.

[Yilmaz 2008a] M. Yilmaz and T. Olsen, A case study of timing-aware ATPG, in *Proceedings Mentor User2User Conference*, 2008.

[Yilmaz 2008b] M. Yilmaz, K. Chakrabarty, and M. Tehranipoor, Interconnect-aware and layout-oriented test-pattern selection for small-delay defects, in *Proceedings International Test Conference [ITC'08]*, 2008.

[Yilmaz 2008c] M. Yilmaz, K. Chakrabarty, and M. Tehranipoor, Test-pattern grading and pattern selection for small-delay defects, in *Proceedings IEEE VLSI Test Symposium*, 2008, pp. 233–239.

10

Conclusion

Continuous advancement in semiconductor process technology enables the creation of large and complex system chips. On one hand, the use of advanced process technology helps meet modern electronic industry needs regarding computing processing power, speed, die size, and form factor; on the other hand, it also brings forward several test challenges. Some of these test challenges are associated with the test data volume and test pattern size due to the sheer large size of the chip; other test challenges are new and not well understood. One such critical challenge is testing and diagnosis of small-delay defects. Small-delay defects can cause immediate failure of a circuit if introduced on critical paths, whereas they cause major quality concerns if they occur on noncritical paths. It is important to note that small-delay defects are not a new phenomenon in the semiconductor industry; they existed as well in the older process nodes. However, their importance and impact on product quality have increased significantly in advanced nodes. In today's multigigahertz and multimillion transistor designs, testing for small-delay defects is critical to ensure product quality during the short life span of the product.

Traditional test methods such as stuck-at fault testing and transition fault testing do provide good coverage of most defects. However, where stuck-at fault testing does not cover small-delay defects, the transition fault testing is unable to target and detect all sorts of small-delay defects. For example, small-delay defects on very short paths are most likely to be untested by the traditional transition fault patterns since traditional transition fault patterns are usually generated along the easiest testable path, which can be the shortest, medium-length, or longest path. Path delay patterns can provide good coverage of small-delay defects, but they suffer from an exponential increase in the number of paths as the design complexity increases. In short, high-quality testing of small-delay defects requires special (usually dedicated) test methods that can be classified into three basic categories: (1) timing-aware pattern generation, (2) faster-than-at-speed tests, and (3) pattern/fault selection-based hybrid techniques. This book presented detailed analyses of various test methods used in the industry as well as proposed by academia for testing of small-delay defects.

Chapter 1 presented basics about various fault models, test methods, and design-for-test (DfT) techniques. A detailed overview of transition fault testing (launch on shift and launch on capture) and test pattern generation (robust and nonrobust) was presented. This overview is essential in understanding how transition fault patterns help in detecting small-delay defects. Next, the need for special testing for small-delay defects in modern advance process was established.

Chapters 2 and 3 presented timing-aware pattern generation techniques for testing of small-delay defects. In timing-aware patterns, the timing information from the static timing analysis (STA) of the design is taken into account while generating the patterns for each fault site. During pattern generation for each fault site, the ATPG (automatic test pattern generation) tool tries to take the longest testable path for fault excitation and propagation to maximize the detection of any small-delay defect. In the test method based on the K longest path as described in Chapter 2, test patterns are generated by enumeration and calculating the K longest path for each fault site; the timing-aware pattern generation described in Chapter 3 is more classical and considers the longest testable path for each fault site only. Some of the major drawbacks of timing-aware ATPG are large test pattern count and long computation time.

Test methods based on a faster-than-at-speed approach were described in Chapters 4 and 5. In the faster-than-at-speed testing, existing or modified transition fault patterns were applied at a clock speed higher than the system clock. The application of a test pattern at a higher clock reduces the available system slack on all paths, thereby increasing the chances of detection of small-delay defects on all paths. Due to this, test methods based on a faster-than-at-speed approach not only are able to detect critical small-delay defects but also can screen out possible reliability defects. However, special care should be taken when selecting the test clock speed as many of the long paths may become multicycle paths and hence should be masked. Also, a higher test frequency will expose more short paths to possible delay defects and hence could result in yield overkill for products that do not require stringent quality levels. Nevertheless, the problem of a large pattern count remains a major disadvantage of faster-than-at-speed test methods. The faster-than-at-speed-based test methods were also extended to include the effects of process variation and crosstalk effects in Chapter 5. This extension is critical for understanding small-delay defects since they not only are caused by the physical defects introduced during manufacturing but also happen in real time due to electrical defects such as cross talk.

Alternative or hybrid test methods that resolve the issue of the large pattern count and run time without compromising the test quality were described in Chapters 6, 7, and 8. Most of the alternative test methods are based on either effective test pattern selection or effective fault site selection. In pattern selection-based methods as described in Chapter 6, instead of generating timing-aware patterns, effective test patterns are selected from an existing pool of transition fault patterns. The effectiveness of the test patterns toward small-delay defects is measured based on the output deviation observed at primary/scan outputs in the circuits. Similarly, other metrics can be used to find test patterns that have higher coverage of small-delay defects. For large industrial circuits, these methods provide greater flexibility in terms of test quality and test cost. However, one of the drawbacks of these methods is that the overall quality of results cannot be better than the quality of the pool of test patterns from which the test patterns are selected. Therefore, having a very good quality baseline test pattern set is a prerequisite for these methods. Effective fault selection-based alternative methods do not suffer from this drawback. The main motivation behind effective fault-selected based methods is that not all fault sites require special test patterns to target small-delay defects as the traditional transition fault patterns already detect a large number of possible small-delay defects. The fault simulation-based fault selection approach was described in Chapter 7; Chapter 8 described a fault selection method based on circuit topology. Both approaches result in a good-quality test set within a reasonable computation time. Industrial uses of both approaches were also presented to validate the worthiness of the approaches for production use.

Finally, a detailed analysis of different test quality metrics used to measure the small-delay defect effectiveness for a give pattern set was presented in Chapter 9. Not only existing metrics were explored for different usage (product quality vs. test quality), but also a new metric was proposed. The baseline assumption behind the new metric is that test quality should change with the test pattern set and should have better correlation with actual coverage and size of possibly detectable small-delay defects.

In terms of future research challenges, effective diagnosis and physical failure analysis (PFA) of failures related to small-delay defects are critical for understanding the behavior and sources of these defects. Considering that diagnosis/PFA of transition faults is tedious and becoming challenging in advanced process nodes, diagnosis/PFA of small-delay defects are nothing less than formidable. In terms of the future, as the technology moves beyond 20 nm and special circuit structures like FinFet are used, a clear and through understanding of small-delay defects may be required. It will be interesting to see how small-delay defects affect the design and performance of circuits in the case of FinFet as compared to bulk complementary metal oxide semiconductors (CMOSs), which has been the field of study since 1975 or the early eighties.

References

[Abramovici 1990] M. Abramovici, M.A. Breuer, and A.D. Friedman, *Digital Systems Testing and Testable Design*, Wiley-IEEE Press, New York, 1990.

[Acken 1992] J.M. Acken and S.D. Millman, Fault model evolution for diagnosis accuracy versus precision, in *Proceedings Custom Integrated Circuits Conference*, 1992, pp. 13.4.1–13.4.4.

[Ahmed 2005] N. Ahmed, C. Ravikumar, M. Tehranipoor, and J. Plusquellic, *At-Speed Transition Fault Testing with Low Speed Scan Enable*, IEEE Computer Society, 2005.

[Ahmed 2006a] N. Ahmed, M. Tehranipoor, and V. Jayaram, A novel framework for faster-than-at-speed delay test considering IR-drop effects, in *Proceeding International Conference on Computer-Aided Design*, 2006.

[Ahmed 2006b] N. Ahmed, M. Tehranipoor, and V. Jayram, Timing-based delay test for screening small delay defects, in Proceedings IEEE Design Automation Conference, 2006.

[Ahmed 2007] N. Ahmed, M. Tehranipoor, and V. Jayaram, Supply voltage noise aware ATPG for transition delay faults, in *Proceedings IEEE VLSI Test Symposium*, 2007.

[Ahmed 2010] N. Ahmed and M. Tehranipoor, A novel faster-than-at-speed transition delay test method considering IR-drop effects, in *IEEE Transactions on Computer-Aided Design* 28, pp. 1573–1582, 2010.

[Allampally 2005] S. Allampally, V. Prasanth, R.K. Tiwari, and K.S., Small delay defect testing in DSM technologies—Needs and considerations, Synopsys Users Group Conference, 2005.

[Amyeen 2004] M.E. Amyeen, S. Venkataraman, A. Ojha, and S. Lee, Evaluation of the quality of N-detect scan ATPG patterns on a processor, in *Proceedings IEEE International Test Conference [ITC'04]*, 2004, pp. 669–678.

[Barnhart 2004] C. Barnhart, What is true-time delay test? *Cadence Nanometer Test Q* 1[2], 2004.

[Bell 1996] J.A. Bell, Timing Analysis of Logic-Level Digital Circuits Using Uncertainty Intervals, MS thesis, Department of Computer Science, Texas A&M University, 1996.

[Benkoski 1990] J. Benkoski, E.V. Meersch, L.J.M. Claesen, and H.D. Man, Timing verification using statically sensitizable paths, *IEEE Transactions on Computer-Aided Design*, 9(10), pp. 1073–1084, October 1990.

[Benware 2003] B. Benware, C. Schuermyer, N. Tamarapalli, K.-H. Tsai, S. Ranganathan, R. Madge, J. Rajski, and P. Krishnamurthy, Impact of multiple-detect test patterns on product quality, *Proceedings International Test Conference*, 2003, pp. 1031–1040.

[Brand 1994] D. Brand and V.S Iyengar, Identification of redundant delay faults, *IEEE Transactions on CAD* of Integrated Circuits and Systems, 13(5), pp. 553–565, 1994.

[Braun 2005] A.E. Braun, Pattern-related defects become sublter deadlier, *Semiconductor International*, June 1, 2005.

[Bushnell 2000] M.L. Bushnell and V.D. Agrawal, *Essentials of Electronic Testing for Digital, Memory and Mixed-Signal VLSI Circuits*, Springer, New York, 2000.

[Butler 2008] K.M. Butler, J.M Carulli, and J. Saxena, Modeling test escape rate as a function of multiple coverages, in *Proceedings International Test Conference*, 2008.

[Cadence 2004] Cadence Inc., *User Manual for Cadence Encounter Toolset Version 2004.10*, Cadence Inc., San Jose, CA, 2004.

[Cadence 2005] Cadence Inc., *0.18 μm Standard Cell GSCLib Library Version 2.0*, Cadence Inc., San Jose, CA, 2005.

[Cadence 2006] Cadence Inc., *User Manual for Encounter True-time Test ATPG*, Cadence Inc., San Jose, CA, 2006.

[Carter, 1987] J.L. Carter, V.S. Iyengar, and B.K. Rosen, Efficient test coverage determination for delay faults, in *Proceedings IEEE International Test Conference*, 1987, pp. 418–427.

[Chang 1993] H. Chang and J.A. Abraham, VIPER An efficient vigorously sensitizable path extractor, in *ACM/IEEE Design Automation Conference*, 1993, pp. 112–117.

[Chang 2008] C.-J. Chang and T. Kobayashi, Test quality improvement with timing-aware ATPG: Screening small delay defect case study, poster session, International Test Conference, 2008.

[Chao 2004] C.-T.M. Chao, L.-C. Wang, and K.-T. Cheng, Pattern selection for testing of deep sub-micron timing defects, in *Proceedings IEEE Design Automation and Test in Europe*, 2004, vol. 2, pp. 1060–1065.

[Chen 1997] W. Chen, S. Gupta, and M. Breuer, Analytic models for crosstalk delay and pulse analysis under non-ideal inputs, in *Proceedings IEEE International Test Conference*, 1997, pp. 808–818.

[Cheng 1996] K.-T. Cheng and H.-C. Chen, Classification and identification of non-robustly untestable path delay faults, *IEEE Transactions on Computer-Aided Design Integrated Circuits and Systems* 15(8), pp. 845–853, August 1996.

[Cheng 2000] K.-T. Cheng, S. Dey, M. Rodgers, and K. Roy, Test challenges for deep sub-micron technologies, in *Proceedings IEEE Design Automation Conference*, 2000, pp. 142–149.

[Daasch 2005] W.R. Daasch and Robert Madge, Data-driven models for statistical testing: Measurements, estimates and residuals, in *Proceedings IEEE International Conference*, 2005, p. 10.

[de Sousa 1994] J.T. de Sousa, F.M. Goncalves, J.P. Teixeira and T.W. Williams Fault modeling and defect level projections in digital ICs, in *Proceedings European Design and Test Conference (EDTC)*, 1994, pp. 436–442.

[de Sousa 2000] J.T. de Sousa and V.D. Agrawal, Reducing the complexity of defect level modeling using the clustering effect, in *Proceedings Design, Automation and Test in Europe Conference and Exhibition*, 2000, pp. 640–644.

[Devta-Prasanna 2009] N. Devta-Prasanna, S.K. Goel, A. Gunda, M. Ward, and P. Krishnamurthy, Accurate measurement of small delay defect coverage of test patterns, in *Proceedings International Test Conference*, 2009.

[Dumas 1993] D. Dumas, P. Girard, C. Landrault, and S. Pravossoudovitch, An implicit delay fault simulation method with approximate detection threshold calculation, in *Proceedings International Test Conference*, 1993, pp.705–713.

[Ferhani 2006] F.-F. Ferhani and E. McCluskey, Classifying bad chips and ordering test sets, in *Proceedings IEEE International Test Conference*, 2006.

[Forzan 2007] C. Forzan and D. Pandini, Why we need statistical static timing analysis, in *Proceedings IEEE International Conference on Computer Design*, 2007, pp. 91–96.

[Foster 1976] R. Foster, Why consider screening, burn-in, and 100-percent testing for commercial devices? *IEEE Transactions on Manufacturing Technology* 5(3), pp. 52–58, September 1976.

[Fuchs 1991] K. Fuchs, F. Fink, and M.H. Schulz, DYNAMITE: An efficient automatic test pattern generation system for path delay faults, *IEEE Transactions on Computer-Aided Design* 10(10), pp. 1323–1355, October 1991.

[Fuchs 1994] K. Fuchs, M. Pabst, and T. Rossel, RESIST: A recursive test pattern generation algorithm for path delay faults considering various test classes, *IEEE Transactions on Computer-Aided Design* 13(12), pp. 1550–1562, December 1994.

[Gaisler 2006] Gaisler Research, 2006, http://gaisler.com

[Geuzebroek 2007] J. Geuzebroek, E.J. Marinissen, A.K. Majhi, A. Glowatz, and F. Hapke, Embedded multi-detect ATPG and its effect on the detection of unmodeled defects, in *Proceedings International Test Conference*, 2007, pp. 1–10.

[Goel 1981] P. Goel, An implicit enumeration algorithm to generate tests for combinational logic circuits, *IEEE Transactions on Computers* C-30(3), pp. 215–222, March 1981.

[Goel 2009] S.K. Goel, N. Devta-Prasanna, and R. Turakhia, Effective and efficient test pattern generation for small delay defects, in *Proceedings IEEE VLSI Test Symposium*, 2009.

[Goel 2010] S.K. Goel, K. Chakrabarty, M. Yilmaz, K. Peng, and M. Tehranipoor, Circuit topology-based test pattern generation for small delay defects, in *Proceedings IEEE Asian Test Symposium*, 2010.

[Gupta 2004] P. Gupta and M. Hsiao, ALAPTF: A new transition fault model and the ATPG algorithm, in *Proceedings IEEE International Test Conference*, 2004, pp. 1053–1060.

[Hansen 1999] M. Hansen, H. Yalcin, and J.P. Hayes, Unveiling the ISCAS-85 benchmarks: A case study in reverse engineering, *IEEE Design and Test of Computers* 16(3), pp. 72–80, July–September 1999.

[Hao 1993] H. Hao and E. McCluskey, Very-low-voltage testing for weak CMOS logic ICs, in *Proceedings IEEE International Test Conference*, 1993, pp. 275–284.

[Heragu, 1996] K. Heragu, J.H. Patel, and V.D. Agrawal, Segment delay faults: A new fault model, in *Proceedings IEEE VLSI Test Symposium*, 1996, pp. 32–39.

[Hess 2010] C. Hess, A. Inani, A. Joag, Zhao Sa, M. Spinelli, M.Zaragoza, Nguyen Long, and B. Kumar, Stackable short floe characterization vehicle test chip to reduce test chip designs, mask cost and engineering wafers, in *Proceedings Advanced Semiconductor Manufacturing Conference*, 2010.

[IEEE 1497] Institute of Electrical and Electronics Engineers, *IEEE Standard for Standard Delay Format [SDF] for the Electronic Design Process*, http://www.ieeexplore.ieee.org/iel5/7671/20967/00972829.pdf

[ITRS 2007] International Technology Roadmap for Semiconductors, International Technology Roadmap for Semiconductor Industry, 2007, http://www.itrs.net/Links/2007ITRS/Home2007.htm

[ITRS 2008] International Technology Roadmap for Semiconductors, International Technology Roadmap for Semiconductor Industry, 2008, http://www.itrs.net/Links/2008ITRS/Home2008.htm

[IWLS 2005] International Workshop on Logic and Synthesis, IWLS Benchmarks, 2005, http://iwls.org/iwls2005/benchmarks.html

[Iyengar 1988a] V.S. Iyengar, B.K. Rosen, and I. Spillinger, Delay test generation I—Concepts and coverage metrics, in *Proceedings IEEE International Test Conference*, 1988, pp. 857–866.

[Iyengar 1988b] V.S. Iyengar, B.K. Rosen, and I. Spillinger, Delay test generation II—Concepts and coverage metrics, in *Proceedings IEEE International Test Conference*, 1988, pp. 867 876.

[Jha 2003] N.K. Jha and S.K. Gupta, *Testing of Digital Systems*, Cambridge University Press, Cambridge, UK, 2003.

[Kajihara 2007] S. Kajihara, S. Morishima, M. Yamamoto, X. Wen, M. Fukunaga, K. Hatayama, and T. Aikyo, Estimation of delay test quality and its application to test generation, in *Proceedings IEEE International Conference on Computer-Aided Design*, 2007.

[Kapur 2007] R. Kapur, J. Zejda, and T.W. Williams, Fundamental of timing information for test: How simple can we get?, in *Proceedings IEEE International Test Conference*, 2007.

[Keller 2004] B. Keller, M. Tegethoff, T. Bartenstein, and V. Chickermane, An economic analysis and ROI model for nanometer test, in *Proceedings IEEE International Test Conference*, 2004, pp. 518–524.

[Killpack 2008] K. Killpack, S. Natarajan, A. Krishnamachary, and P. Bastani, Case study on speed failure causes in a microprocessor, *IEEE Design and Test of Computers* 25(3), pp. 224–230, May–June 2008.

[Kim 2003] K.S. Kim, S. Mitra, and P.G. Ryan, Delay defect characteristics and testing strategies, *IEEE Design and Test of Computers* 20(5), pp. 8–16, September–October 2003.

[Konuk 2000] H. Konuk, On invalidation mechanisms for non-robust delay tests, in *Proceedings International Test Conference*, 2000 pp. 393–399.

[Koralov 2007] L.B. Koralov and Y.G. Sinai, *Theory of Probability and Random Processes*, 2nd edition, Springer, New York, 2007.

[Krishnaswamy 2001] V. Krishnaswamy, A.B. Ma, and P. Vishakantaiah, A study of bridging defect probabilities on a Pentium™ 4 CPU, in *Proceedings International Test Conference*, 2001, pp. 688–695.

[Krstic 1998] A. Krstic and K.-T. Cheng, *Delay Fault Testing for VLSI Circuits*, Springer, Boston, 1998.

[Kruseman 2004] B. Kruseman, A.K. Majhi, G. Gronthoud, and S. Eichenberger, On hazard-free patterns for fine-delay fault testing, in *Proceedings IEEE International Test Conference*, 2004.

[Kuhn 2008] K. Kuch, C. Kenyon, A. Kornfield, M. Liu, A. Maheshwari, W.-K. Shih, S. Sivakumar, G. Taylor, P. VanDerVoorn, and K. Zawadzki, Managing process variation in Intel's 45nm CMOS technology, *Intel Technology Journal* 2(17), 2008.

[Kulkarni 2006] S.H. Kulkarni, D. Sylvester, and D. Blaauw, A statistical framework for post-silicon tuning through body bias clustering, in *Proceedings of IEEE/ACM International Conference on Computer-Aided Design (ICCAD)*, 2006, pp. 39–46.

[Lee 2005] B.N. Lee, L.-C. Wang, and M. Abadir, Reducing pattern delay variations for screening frequency dependent defects, in *Proceedings IEEE VLSI Test Symposium*, 2005, pp. 153–160.

[Lee 2006] H. Lee, S. Natarajan, S. Patil, and I. Pomeranz, Selecting high-quality delay tests for manufacturing test and debug, in *Proceedings IEEE International Symposium on Defect and Fault-Tolerance in VLSI Systems*, 2006.

[Li 1989] W.N. Li, S.M. Reddy and S.K. Sahni, On path selection in combinational logic circuits, *IEEE Transactions on Computer-Aided Design* 8(1), pp. 56–63, January 1989.

[Li 2000] H. Li, Z. Li, and Y. Min, Reduction in number of path to be tested in delay testing, *Journal of Electronic Testing: Theory and Application [JETTA]*, 16(5), pp. 477–485, 2000.

[Lin 1987] C.J. Lin and S.M. Reddy, On delay fault testing in logic circuits, *IEEE Transactions on Computer-Aided Design*, 6(9), pp. 694–701, September 1987.

[Lin 2001] X. Lin, J. Rajski, I. Pomeranz, and S.M. Reddy, On static test compaction and test pattern ordering for scan design, in *Proceedings IEEE International Test Conference*, 2001, pp. 1088–1097.

[Lin 2003] X. Lin, R. Press, J. Rajski, P. Reuter, T. Rinderknecht, B. Swanson, and N. Tamarapalli, High frequency, at-speed scan testing, *IEEE Design and Test Computers* 20(5), pp. 17–25, September–October 2003.

[Lin 2006] X. Lin, K.-H. Tsai, C. Wang, M. Kassab, J. Rajski, T. Kobayashi, R. Klingenberg, Y. Sato, S. Hamada, and T. Aikyo, Timing-aware ATPG for high quality at-speed testing of small delay defects, in *Proceedings IEEE Asian Test Symposium*, 2006, pp. 139–146.

[Lin 2007a] X. Lin, M. Kassab, and J. Rajski, Test generation for timing-critical transition faults, in *Proceedings IEEEE Asian Test Symposium*, 2007, pp. 493–500.

[Lin 2007b] X. Lin, K.-H. Tsai, M. Kassab, C. Wang, and J. Rajski, Timing-aware test generation and fault simulation, U.S. Patent Application 20070288822, December 2007.

[Lion 2002] J. Lion, A. Krstic, L. Wang, and K. Cheng, False-path-aware statistical timing analysis and efficient path selection for delay testing and timing validation, in *Proceedings IEEE Design Automation Conference*, 2002, pp. 566–569.

[Liou 2003] J.-J. Liou, L.-C. Wang, A. Krstic, and K.-T. Cheng, Experience in critical path selection for deep submicron delay test and timing validation, in *Proceedings of ASP-DAC*, 2003, pp. 751–756.

[Mahmoodi 2005] H. Mahmoodi, S. Mukhopadhyay, and K. Roy, Estimation of delay variations due to random-dopant fluctuations in nanoscale CMOS circuits, *IEEE Journal of Solid-State Circuits* 40(9), pp. 1787–1796, September 2005.

[Majhi 2000] A.K. Majhi, V.D. Agrawal, J. Jacob, and L.M. Patnaik, Line coverage of path delay faults, *IEEE Transactions on Very Large Scale Integration [VLSI] Systems* 8(5), pp. 610–614, 2000.

[Mattiuzzo 2008] R. Mattiuzzo, D. Appello, and C. Allsup. Small delay defect testing, in *Proceedings Synopsys User's Group Conference*, 2008.

[Mattiuzzo 2009] R. Mattiuzzo, D. Appello, and C. Allsup, Small delay defect testing, *Test and Measurement World*, 2009.

[Maxwell 1993] P.C. Maxwell and R.C. Aitken, Biased voting: A method for simulating CMOS bridging faults in the presence of variable gate logic thresholds, in *Proceedings International Test Conference*, 1993, pp. 63–72.

[Maxwell 1994] P.C. Maxwell, R.C. Aitken, and L.M. Huisman, The effect on quality of non-uniform fault coverage and fault probability, *ITC1994*, 1994, pp. 739–746.

[Mei 1974] K. Mei, Bridging and stuck-at faults, *IEEE Transactions on Computers* C-23(7), pp. 720–727 (1974).

[Meijer 2012] M. Meijer and J.P. Gyvez, Body-bias-driven design strategy for area- and performance-efficient CMOS circuits, *IEEE Transactions on Very Large Scale Integration [VLSI] Systems* 20(1), pp. 42–51, 2012.

[Mentor 2006] *Understanding How to Run Timing-Aware ATPG*, Application Note, 2006.

[Mentor 2008] *Scan and ATPG Process Guide, FastScan Tool*, Application Note, August 2008.

[Mitra 2004] S. Mitra, E. Volkerink, E.J. McCluskey, and S. Eichenberger, Delay defect screening using process monitor structures, in *Proceedings IEEE VLSI Test Symposium*, 2004.

[Murakami 2000] A. Murakami, S. Kajihara, T. Sasao, R. Pomeranz, and S.M. Reddy, Selection of potentially testable path delay faults for test generation, in *Proceedings IEEE International Test Conference*, 2000, pp. 376–384.

[Nigh 2000] P. Nigh and A. Gattiker, Test method evaluation experiments and data, in *Proceedings IEEE International Test Conference*, 2000, pp. 454–463.

[Nitta 2007] I. Nitta, S. Toshiyuki, and H. Katsumi, Statistical static timing analysis technology, *Journal of Fujitsu Science and Technology* 43(4), pp. 516–523, October 2007.

[Padmanaban 2004] S. Padmanaban and S. Tragoudas, A critical path selection method for delay testing, in *Proceedings IEEE International Test Conference*, 2004, pp. 232–241.

[Park 1989] E.S. Park, M.R. Mercer, and T.W. Williams, A statistical model for delay-fault testing, *IEEE Design and Test of Computers* 6(1), pp. 45–55, 1989.

[Park 1991] E.S. Park, B. Underwood, T.W. Williams, and M. Mercer, Delay testing quality in timing-optimized designs, in *Proceedings International Test Conference*, October 1991, pp. 897–905.

[Park 1992] E.S. Park, M.R. Mercer, and T.W. Williams, The total delay fault model and statistical delay fault coverage, in *IEEE Transactions on Computers* 41(6), pp. 688–698, June 1992.

[Patil 1992] S. Patil and J. Savir, Skewed-load transition test: Part II, coverage, in *Proceedings IEEE International Test Conference*, 1992, pp. 714–722.

[Pomeranz 1995] I. Pomeranz, S.M. Reddy, and P. Uppaluri, NEST: A nonenumerative test generation method for path delay faults in combinational circuits, *IEEE Transactions on Computer-Aided Design* 14(12), pp. 1505–1515, December 1995.

[Pomeranz 1998] I. Pomeranz and S.M. Reddy, A generalized test generation procedure for path delay faults, in *Proceedings International Symposium on Fault-Tolerant Computing*, 1998, pp. 274–283.

[Pomeranz 1999] I. Pomeranz and S.M. Reddy, On n-detection test sets and variable n-detection test sets for transition faults, in *Proceedings IEEE VLSI Test Symposium*, 1999, pp. 173–180.

[Pomeranz 2008] I. Pomeranz and S.M. Reddy, Transition path delay faults: A new path delay fault model for small and large delay defects, *IEEE Transactions on Very Large Scale Integration [VLSI] Systems* 16(1), pp. 98–107, January 2008.

[Pomeranz 2010] I. Pomeranz and S.M. Reddy, Hazard-based detection conditions for improved transition path delay fault coverage, *IEEE Transactions on CAD of Integrated Circuits and Systems* 29(9), pp. 1449–1453, 2010.

[Pramanick 1997] A.K. Pramanick and S.M. Reddy, On the fault detection coverage of gate delay fault detecting tests, *IEEE Transactions on Computer-Aided Design* 16(1), pp. 78–94, 1997.

[Putman 2006] R. Putman and R. Gawde, Enhanced timing-based transition delay testing for small delay defects, in *Proceedings IEEE VLSI Test Symposium*, 2006.

[Qiu 2003] W. Qiu and D.M.H. Walker, An efficient algorithm for finding the K longest testable paths through each gate in a combinational circuit, in *Proceedings IEEE International Test Conference*, 2003, pp. 592–601.

[Qiu 2004] W. Qiu, J. Wang, D. Walker, D. Reddy, X. Lu, Z. Li, W. Shi, and H. Balachandran, K longest paths per gate [KLPG] test generation for scan-based sequential circuits, in *Proceedings IEEE International Test Conference*, 2004, pp. 223–231.

[Rabaey 2003] J.M. Rabaey, A. Chandrakasan, and B. Nikolic, *Digital Integrated Circuits, a Design Perspective*, 2nd edition, Prentice Hall, Upper Saddle River, NJ, 2003.

[Raina 2006] R. Raina, What is DFM and DFY and why should i care? in *Proceedings International Test Conference*, 2006, pp. 1–9.

[Recktenwald 2000] G.W. Recktenwald, *Numerical Methods with MATLAB: Implementations and Applications*, Prentice Hall, Upper Saddle River, NJ, 2000.

[Reddy 1996] S.M. Reddy, I. Pomeranz, and S. Kajihara, On the effects of test compaction on defect coverage, in *Proceedings VLSI Test Symposium*, 1996.

[Sato 2005a] Y. Sato, S. Hamada, T. Maeda, A. Takatori, and S. Kajihara, Evaluation of the statistical delay quality model, in *Proceedings of Asia and South Pacific Design Automation Conference*, 2005, pp. 305–310.

[Sato 2005b] Y. Sato, S. Hamada, T. Maeda, A. Takatori, and S. Kajihara, Invisible delay quality—SDQM model lights up what could not be seen, in *Proceedings International Test Conference*, 2005, pp. 1202–1210.

[Savir 1992] J. Savir, Skewed-load transition test: Part I, calculus, in *Proceedings IEEE International Test Conference*, 1992, pp. 705–713.

[Savir 1994a] J. Savir and S. Patil, Broad-side delay test, *IEEE Transactions on Computer-Aided Design* 13(8), pp. 1057–1064, August 1994.

[Savir 1994b] J. Savir and S. Patil, On broad-side delay test, in *Proceedings IEEE VLSI Test Symposium*, 1994, pp. 284–290.

[Saxena 2002] J. Saxena, K. Butler, J. Gatt, R. Raghuraman, S. Kumar, S. Basu, D. Campbell, and J. Berech, Scan-based transition fault testing—implementation and low cost test challenges, in *Proceedings IEEE International Test Conference*, 2002, pp. 1120–1129.

[Segura 2002] J. Segura, A. Keshavarzi, J.M. Soden, and C.F. Hawkins, Parametric failures in CMOS ICs—A defect-based analysis, in *Proceedings International Test Conference*, 2002, pp. 90–99.

[Sengupta 1999] S. Sengupta et al., Defect-based tests: A key enabler for successful migration to structural test, *Intel Technology Journal* Quarter 1, 1999.

[Shao 2002] Y. Shao, S.M. Reddy, I. Pomeranz, and S. Kajihara, On selecting testable paths in scan designs, in *Proceedings IEEE European Test Workshop*, Corfu, Greece, 2002, pp. 53–58.

[Sharma 2002] M. Sharma and J.H. Patel, Finding a small set of longest testable paths that cover every gate, in *Proceedings IEEE International Test Conference*, 2002, pp. 974–982.

[Smith 1985] G.L. Smith, Model for delay faults based upon paths, in *Proceedings IEEE International Test Conference*, 1985, pp. 342–349.

[Smith 1997] M. Smith, *Application-Specific Integrated Circuits*, Addison-Wesley Professional, Reading, MA, 1997.

[Srivastava 2005] A. Srivastava, S. Shah, K. Agrawal, D. Sylvester, D. Blaauw, and S. Director, Accurate and efficient gate-level parametric yield estimation considering correlated variations in leakage power and performance, in *Proceedings of Design Automation Conference*, 2005.

[Stewart 1991] R. Stewart and J. Benkoski, Static timing analysis using interval constraints, in *Proceedings IEEE International Conference on Computer-Aided Design*, 1991, pp. 308–311.

[Synopsys 2007a] Synopsys Inc., *SOLD Y-2007*, Volumes 1–3, Synopsys Inc., Mountain View, CA, October 2007.

[Synopsys 2007b] Synopsys Inc., *User Manual for Synopsys Toolset Version 2007.03*, Synopsys Inc., Mountain View, CA, 2007.

[Synopsys 2008] Synopsys Inc., *TetraMax ATPG User Guide*, Synopsys Inc., Mountain View, CA, 2008.

[Tendolkar 1985] N. Tendolkar, Analysis of timing failures due to random AC defects in VLSI modules, in *Proceedings IEEE Design Automation Conference*, 1985.

[Trivedi 2001] K. Trivedi, *Probability and Statistics with Reliability, Queuing, and Computer Science Applications*, 2nd edition, Wiley, New York, 2001.

[Turakhia 2007] R. Turakhia, W.R. Daasch, M. Ward, and J. van Slyke, Silicon evaluation of longest path avoidance testing for small delay defects, in *Proceedings IEEE International Test Conference*, 2007.

[Uzzaman 2006] A. Uzzaman, M. Tegethoff, B. Li, K. McCauley, S. Hamada, and Y. Sato, Not all delay tests are the same—SDQL model shows true-time, in *Proceedings IEEE Asian Test Symposium*, 2006, pp. 147–152.

[Waicukauski 1987] J.A. Waicukauski, E. Lindbloom, B.K. Rosen, and V.S. Iyengar, Transition fault simulation, *IEEE Design and Test of Computer* 4(2), pp. 32–38, April 1987.

[Wang 2008a] Z. Wang and K. Chakrabarty, Test-quality/cost optimization using output-deviation-based reordering of test patterns, *IEEE Transactions on Computer-Aided Design of Integrated Circuits and Systems* 27, pp. 352–365, February 2008.

[Wang 2008b] Z. Wang and D.M.H. Walker, Dynamic compaction for high quality delay test, in *Proceedings IEEE VLSI Test Symposium*, 2008, pp. 243–248.

[Wang 2009] Z. Wang and D.M.H. Walker, Compact delay test generation with a realistic low cost fault coverage metric, in *Proceedings IEEE VLSI Test Symposium*, 2009, pp. 59–64.

[Williams 1981] T.W. Williams and N.C. Brown, Defect level as a function of fault coverage, *IEEE Transactions on Computers* C-30, pp. 987–988, 1981.

[Williams 1991] T.W. Williams, B. Underwood, and M.R. Mercer, The interdependence between delay-optimization of synthesized networks and testing, in *Proceedings ACM/IEEE Design Automation Conference*, 1991, pp. 87–92.

[Xie 2009] Y. Xie and Y. Chen, Statistical high level synthesis considering process variations, *IEEE Computer Design and Test*, Special Issue on HLS, 26(4), pp. 78–87, July–August, 2009.

[Yan 2004] H. Yan and A.D. Singh, Evaluating the effectiveness of detecting delay defects in the slack interval: A simulation study, in *Proceedings IEEE International Test Conference*, 2004.

[Yilmaz 2008a] M. Yilmaz and T. Olsen, A case study of timing-aware ATPG, in *Proceedings Mentor User2User Conference*, 2008.

[Yilmaz 2008b] M. Yilmaz, K. Chakrabarty, and M. Tehranipoor, Interconnect-aware and layout-oriented test-pattern selection for small-delay defects, in *Proceedings International Test Conference [ITC'08]*, 2008.

[Yilmaz 2008c] M. Yilmaz, K. Chakrabarty, and M. Tehranipoor, Test-pattern grading and pattern selection for small-delay defects, in *Proceedings IEEE VLSI Test Symposium*, 2008, pp. 233–239.

[Yilmaz 2010] M. Yilmaz, K. Chakrabarty, and M. Tehranipoor, Test-pattern selection for screening small-delay defects in very-deep sub-micrometer integrated circuits, in *IEEE Transactions on Computer-Aided Design of Integrated Circuits and Systems* 29(5), pp. 760–773, 2010.

[Zolotov 2010] V. Zolotov, J. Xiong, H. Fatemi, and C. Visweswariah, Statistical path selection for at-speed test, *IEEE Transactions on Computer-Aided Design of Integrated Circuits and Systems* 29(5), pp. 749–759, May 2010.

Index

Printed and bound by CPI Group (UK) Ltd, Croydon, CR0 4YY

18/10/2024

01776269-0001